Lecture Notes in Control and Information Sciences

Edited by A. V. Balakrishnan and M. Thoma

Lecture Notes in Control and Information Sciences

Edited by A.V. Balakrishnan and M. Thoma

10

Jan M. Maciejowski

The Modelling of Systems with Small Observation Sets

Springer-Verlag
Berlin Heidelberg GmbH 1978

Author

Dr. Jan Marian Maciejowski
Maudsley Research Fellow, Pembroke College, Cambridge
also with the
Control and Management Systems Group,
Cambridge University Engineering Department
Mill Lane, Cambridge CB2 1RX, England

ISBN 978-3-540-09004-5   ISBN 978-3-540-35675-2 (eBook)
DOI 10.1007/978-3-540-35675-2

© by Springer-Verlag Berlin Heidelberg 1978
Originally published by Springer-Verlag Berlin Heidelberg New York in 1978.

2061/3020-543210

SUMMARY

The problem of assessing and interpreting models of
systems, when only small sets of observations are available,
is introduced and discussed.   System identification is
defined as the progression from a set of observations of
the behaviour of a system, to a theory which accounts for
that behaviour.   The concepts of algorithmic information
theory are drawn on to develop a characterisation of modelling,
which constitutes a partial solution to the problem of system
identification, while taking account of the size of the set
of available observations.   A model is defined to be an
algorithm for computing the output observation set of a system
under specified restrictions.

A general criterion of the quality of a model, its
"information gain", is proposed, and its consistency with
more conventional criteria is discussed.   It is proved that
no "universal modelling algorithm" can exist, in the sense
that it is not possible, in general, to find the model with
the highest information gain.

Information gain is a suitable criterion for a wide class
of models, including nonlinear dynamical stochastic models,
and its computation is straightforward.   The use of information
gain for the assessment of rival models is demonstrated.

The calculation of information gain requires that the model
be expressed as a computer program.   The choice of programming
language is associated with the modeller's a priori beliefs
about the system.   It is shown that this choice becomes
insignificant as the observation sets become large.   A detailed

investigation shows that it is possible to speak precisely of "the smallest language" required to run a particular program.   A priori knowledge assumed about a system can therefore be considered to be defined by the smallest language required to run the model.

Finally, the effect on model assessment of the manner in which system observations are coded is examined.   It is found that a "safe" coding exists, which often leads to the same assessment as would the use of most other codings.

ACKNOWLEDGEMENTS

The idea of examining modelling in the light of algorithmic information theory is due to Professor A.G.J. MacFarlane. His constant encouragement and enthusiasm, as well as detailed criticism, has been an essential ingredient of this work.

I have also benefited from discussions with many members of the Control and Management Systems Group, of whom Dr.F.P. Kelly, Dr. S.R. Watson and Dr. M.B. Beck deserve special mention. The quotation from Newton in the last chapter was pointed out to me by Dr.A.T.Fuller.

Financial support for this research came from the Science Research Council, and in the final stages from Pembroke College.

Roberta Hill has produced her usual excellent standard of typing, but special thanks are due to her for struggling so successfully through chapter 5.

My wife has asked me not to write one of those embarassing acknowledgements, saying how impossible this research would have been without her constant encouragement and support; consequently I shall leave this to the reader's imagination.

CONTENTS

1.   INTRODUCTION

1.1   Motivation

The areas in which the scientific method has been
demonstrably and spectacularly successful are characterised
by the possibility of performing experiments, or making
observations, more or less freely whenever these are deemed
desirable.   The result of this has been that explicit
consideration of the size of the set of observations from
which a model is hypothesised, and to which a model is
fitted, has been neglected.   Any doubts which arise about
the model can be resolved by further experimentation and
observation.

This pleasant property increasingly disappears as one
enters the domains of complex industrial processes, environmental control systems, management systems, and socio-
economic systems.   The work described here aims to clarify
the relationship between the smallness of the available
observation sets for such systems and the degree of usefulness
of the models obtained for them.

Until recently, the class of models which could be used
in scientific investigations was restricted by a very practical
consideration.   The behaviour of the model had to be
understood, and that understanding could only be obtained from
the theory of the model.   The model was constrained to be
sufficiently simple for theoretical investigation to be

possible.

The availability of the computer has changed this situation radically. It is now possible to investigate the behaviour of a model by simulation, with hardly any theoretical understanding of it. Consequently this constraint on the complexity of useful models has been removed, or at least greatly relaxed. It is now possible to postulate a complicated model structure, to observe its simulated behaviour, and to adjust the details of the model until its simulated behaviour resembles the behaviour of the system being investigated.

When is such a model useful? When does it give any understanding of how the system really works? When can it be used as a reliable guide to how the system will behave in the future? The purpose of the work reported in this thesis is to throw some light on these questions. A further aim is to investigate how rival models of the same system behaviour should be assessed. Most of the thesis is ostensibly concerned with the details of rival model assessment, but it is clear that the ability to distinguish between competing models is intimately connected with the ability to say how good an isolated model is.

Why should a simulation model of the type described above not be useful or reliable? If it reproduces the observed system behaviour, is that not sufficient evidence to indicate the quality of the model? In fact, is it not clear that the better the reproduction of the observed behaviour, the better the model? Our answer is no. The basic reason is the possibility of "overfitting" the model, since its complexity is relatively unconstrained, and it is being checked against a small set of data.

Consider the following simple example. Suppose that two measurements are taken of some variable at two different times, and that we have no other information about it. It is desired to predict the value of the variable at some third time. If a linear variation with time of the variable is proposed, on the basis of the two observations, then it is clear that the only reasonable assessment of confidence in the prediction of the model is nil. The predicted value is no more likely (in an intuitive sense) than any other value. However, if a third measurement is taken which agrees with the prediction of the model, confidence in the model immediately increases. It is now possible to say that values predicted by the model are better predictions, in some sense, than mere guesses. If further measurements are taken, and these also agree with the predictions of the model, then confidence in the model increases very quickly. It never amounts to certainty, of course, but after only ten observations, say, one would have little doubt that the next prediction would be correct (which does not imply that it would be).

The confidence which one is willing to ascribe to this model clearly depends on the difference between the number of observations which it "explains" and the number of observations required to construct the model. If all of the available observations are used to construct the model, then we have no confidence in its predictions. This situation can also be described by saying that if the number of arbitrary decisions that have been made about the model, in order to make it fit the observations, is the same as the number of observations, then we have no confidence in the model.

This point was made succinctly by Poincaré, when he dismissed Jeans' classical explanation of the ultraviolet catastrophe and the specific heat of solids (1):

> "It is obvious that by giving suitable dimensions to the communicating tubes between his reservoirs and giving suitable values to the leaks, Jeans can account for any experimental results whatever. But this is not the role of physical theories. They should not introduce as many arbitrary constants as there are phenomena to be explained; they should establish connections between different experimental facts, and above all they should allow predictions to be made."

On the other hand, the accuracy with which the model reproduces the observed behaviour is clearly significant. If only a slight increase in complexity results in a large increase in accuracy, then in some sense fewer "arbitrary constants" have been added to it than the additional "number of phenomena" which it now explains. What is required for model assessment is some "trade-off" between the complexity of a model and its accuracy. A prerequisite

for this is a measure of complexity which is applicable to
a wide class of models.   A major innovation introduced in
this work is the casting of models in such a form, that
poorness of fit of model behaviour to the observed behaviour
appears as a component of model complexity.   The required
trade-off is thus achieved by assessing model complexity in
a suitable manner.

A more orthodox approach to the problem of model
assessment would be to examine the assessment of models
chosen from a small class, and to postulate some statistical
framework.   It may then be possible to formulate the assessment problem as a statistical decision problem.   This type
of approach has indeed been investigated, even for dynamical
models of the type encountered in control studies (2)(3)(4)
(5).   We do not follow such an approach for the following
reasons.

Any method arrived at from statistical considerations
will be appropriate only for a narrow class of models
(such as linear difference-equation models, for example),
set in a particular (statistical) environment (such as
"observations corrupted by white, Gaussian, additive noise").
Such a method will not be useful if two very different models
are being compared - for example, if the system being
investigated is the behaviour of competing firms in some
market, it may be desired to compare a model based on
Forrester's "Industrial Dynamics" techniques (6) with a model

which uses game theory (7) to explain firms' actions and the
market's responses.

Realistic simulation models often contain nonlinear
elements. When such models are also dynamical, it is
usually extremely difficult to describe the evolution of
the probability distributions of relevant variables (8).
Furthermore, when investigating environmental and socio-
economic systems, the most interesting and important behaviours
often occur under transient conditions. When modelling these,
it may not be appropriate to assume stationariness of relevant
processes. Finally,when few observations of a system are
available, and there is little a priori knowledge about it,
the statistical specification of the system's environment
may itself be very uncertain. In this case little is lost
by not assuming it to be known; in fact, misleading
conclusions may be avoided.

These considerations indicate that it may be more
fruitful to investigate the assessment of models of complex,
poorly understood systems by making as few assumptions as
possible and examining the general situation, rather than
by a painstaking and difficult analysis of each model
structure, as it arises.

1.2   Overview of Approach and Results.

We develop a characterisation of modelling which has
three "components": the system to be modelled, a model of

this system, and a criterion of quality of the model.

The system to be modelled is taken to be defined by a pair of sets of observations of its input and output. Since measurements are always obtained with limited resolution and accuracy, each observation is assumed to be rational. Each set of observations is assumed to be finite.   The system therefore looks like a set of discrete-state, discrete-time measurements.   However, it will become evident that this does not constrain the models of such a system to be of the same category.   It merely reflects the realities of data collection.   A system will be defined in more detail in sec. 1.3.

A model of the system is any algorithm which maps certain subsets of the observations onto the output observations.   This definition is broad enough to admit algorithms which would not normally be of much interest, such as those whose interpretation implies a reversed direction of time, or even a lack of any time ordering.   It also allows algorithms which compute functions defined only on the particular observations obtained.   These are useless for deducing how the system may behave in a new situation (presumably the goal of the modelling exercise), but models of this type will serve as a reference, with respect to which the success of the modelling exercise will be assessed. Any restriction to models of a particular type is accomplished by specifying which subsets of the observations lie in the domain of the algorithm, and which elements of the output observations are to be the corresponding images.

For example, deterministic difference equation models need only map successive blocks of input observations to successive outputs, whereas stochastic predicting models of the Wiener - Kolmogorov or Kalman types must map successive blocks of input and past output observations to successive outputs.

The term "algorithm" may be interpreted as "computer program". Thus we think of models as programs for computing the output observations, and these programs may use the specified subsets of the observations to help them in this task. This viewpoint would be excessively arbitrary, if it were not for the power of Church's Thesis (9), which states that any procedure which satisfies the intuitive notion of an "algorithm" can be expressed in any one of the equivalent formalisations of the theory of algorithms, and hence can be expressed as a computer program.

When the model is written as a computer program in some programming lanaguage, the criterion of quality is taken to be the shortness of that program, as measured by the number of characters in the program. The length of the program is a measure of the number of arbitrary decisions which have been made (relative to the programming language) in constructing the model. Furthermore, a model is required to compute the output observations exactly (to the accuracy with which the observations were originally made). In order

to do this, the model must generate internally those terms
which would conventionally be thought of as "fitting errors".
Since the programming language has a finite number of
terminals, the length of the model increases when these
terms increase.   The criterion of quality thus incorporates
a particular trade-off between complexity and approximation.

The above characterisation of modelling is explained in
more detail in Chapter 3.   Support for it is given in section
2.2.   The essence of this support is that the length of
the shortest program required to compute a sequence displays
properties analogous to the properties of the entropy
associated with a probability space.   In particular, a
long sequence, which requires a maximally long program to
compute it, passes every effective test for randomness
(asymptotically, with probability 1).   This suggests that
the amount by which it is possible to "compress" the program
(model) required to compute a set of observations (system)
represents the amount of information which it has been
possible to extract from the observations.   If the only
model which has been found is one that merely reads out the
observations from a look-up table, then no "compression"
has been achieved, and such a model conveys no information
about the observations.

A consequence of our characterisation is that no
algorithm can exist for finding the best model (according to
the above criterion of quality) of an arbitrary system.

The choice of programming language to be used, for assessing the quality of a model, can be viewed as the specification of "what is to be taken for granted". It should therefore be made in the light of the modeller's *a priori* knowledge about the system, and of the purposes of the modelling exercise. In Chapter 4 this connection is examined more closely. It is shown that, if the observation sets are large enough, then the results of model assessment are independent of the choice of programming language. This can be interpreted to mean that the modeller's *a priori* beliefs become less significant as the set of observations available to him grows.

Nevertheless, the assessment of models of small observation sets is dependent on the modeller's specification of his *a priori* beliefs. Consequently such an assessment cannot be taken to be definitive. However, this is mitigated by the fact that the modeller does not need to choose between mutually exclusive sets of *a priori* beliefs: he can stipulate programming lanaguages which imply a greater or smaller state of knowledge.

Several different models, even when written in the same language, will rarely use exactly the same features of that language. It is therefore questionable whether a comparison of their lengths gives a measure of their complexity relative to the *same* set of assumptions. Chapters 5 and 6 resolve this difficulty. Chapter 5 develops a formal equivalent of "a program makes use of such-and-such facilities of a language". A prerequisite for this is a formal method of defining the semantics of programming languages. One such method is outlined in Appendix A. Chapter 6 then uses the concepts developed in Chapter 5 to specify some conditions under which models may be meaningfully compared. It is demonstrated that model assessment is not much affected if these conditions are not met exactly.

The details of the complexity/approximation trade-off, which is inherent in our proposed method of model assessment, depend on the precise manner in which the observations are coded in the programming language. It is convenient to separate this aspect of the selection of a suitable programming language from those aspects considered in Chapter 4; consequently the coding of observations is discussed in Chapter 7. A distinguished minimal coding is shown to exist, and it is argued that this is a natural coding to use for model assessment.

The modelling of one particular system (Box and Jenkins' gas-furnace data (10)) is used as an example throughout. The rival models considered for this system are very simple and in no way represent the range of possibilities discussed in sec. 1.1. Nevertheless, the considerations raised there apply even to these simple models, as will be seen in Chapter 3. It will become apparent that the assessment method proposed in this thesis is immediately applicable to a much larger class of models.

1.3   System Identification, Realisation and Modelling

Modern developments of systems theory emphasise the notion of a dynamical system as an abstract summary of experimental data (11), (12), (13).   Modelling is concerned with the inference of system behaviour under specified but not yet observed conditions, from observations of past behaviour, under known conditions.   The method by which this is achieved is the postulation of abstract structures for the system, which are compatible with these observations, and the selection, from these candidate structures, of one that is preferred on the basis of some criterion.   The modern view of a system, with its heavy emphasis on observations, is therefore a more natural one to adopt, when discussing modelling, than the older and less useful view of a system as "an interconnection of components".

However, if a system is to be modelled, and the success of the modelling is to be gauged by reference to the observations, then as little abstract structure should be imposed upon it, before modelling has begun, as possible. Consequently, we adopt the following definition:

Definition (1.3.1)

A system S is defined to be an ordered pair of observations, $S = (U, Y)$, where:

(i) $U = (u_1, u_2, \ldots, u_M)$ and $Y = (y_1, y_2, \ldots, y_N)$

are the input and output observation sets respectively;

$u_i = (u_1^i, u_2^i, \ldots, u_{\ell_i}^i)$ and $y_i = (y_1^i, y_2^i, \ldots, y_{m_i}^i)$

are ordered sets of observations carried out at time t_i,

where $t_1, t_2, \ldots, t_N$ is the natural time ordering; $u_j^i \in \{rationals\}$ $\cup \{b\}$ where b (blank) denotes a missing observation; similarly for y_j^i; and

(ii) with the convention that $(b, b, \ldots, b) = b$, if $u_i = b$ then $\ell_i = 0$; if $y_i = b$ then $m_i = 0$; if $u_i \neq b$ then $u_{\ell_i}^i \neq b$; if $y_i \neq b$ then $y_{m_i}^i \neq b$; and $y_N \neq b$.

Conditions (ii) serve only to ensure that adding on a set of blanks (missing observations) does not create a new system. For concreteness, we have specified that u_i, y_i refer to observations made at time t_i, since we are interested primarily in dynamical models. However, this interpretation is not essential. Also, each u_i, y_i could be a multidimensional finite array of observations, rather than a one-dimensional array, without affecting later results.

The input observation set is allowed to be empty, in order to admit devices such as noise generators and oscillators, as systems of the form (b, Y). It has been argued that when stating the general problem of system identification, it should not be necessary to distinguish between input and output(14). The two should be lumped together as a "system behaviour", and the task of system identification should

include the separation of "input" and "output". However,
it seems essential to have a means of distinguishing between
the two cases shown in Fig. 1. The black boxes labelled
"source" and "sink" can be expected to have very different
internal structures - consider the difference between a
signal generator and an earthing point. But any identification
procedure must lead to the same model of both unless the
inputs and outputs are distinguished. It is for this reason
that we have defined a system as an _ordered_ pair of observation
sets. The ordering distinguishes the input observations
from the output observations. Note that the observation
sets U and Y, which define a system, are themselves systems
of the form (b, U)(providing that U $\neq$ b) and (b, Y).

Our definition of "system" may at first seem odd,
especially to those familiar with control theory. In this
field it is conventional to define a system by a set of
equations, and investigate its behaviour by examining
properties of the solutions of these equations. We are
concerned, however, with the reverse process. We assume
that we are aware of the existence of a system because of
its interaction with its environment - in other words, because
we have observations of its behaviour. The goal of modelling
is some concise set of "laws" - such as the set of equations
referred to above - which "explain" this interaction. Hence
we prefer to regard the set of equations as a "model", and
the set of observations as a "system".

The definition of "system" which is proposed above is much cruder than the definitions usually encountered. It is worth stating in full one such definition - that of Kalman, Falb and Arbib (11):

Definition (1.3.2)

A <u>dynamical system</u> (input/output sense) is a composite mathematical concept defined as follows:

(a)   (i)   There is a given time set T, a set of input values U, a set of acceptable input functions $\Omega = \{\omega : T \to U\}$, a set of output values Y, and a set of output functions $\Gamma = \{\gamma : T \to Y\}$.

(ii)   (Direction of time). T is an ordered subset of the reals.

(iii) The input space Ω satisfies the following conditions:

    (1) (Nontriviality). Ω is nonempty.

    (2) (Concatenation of inputs).   An input segment $\omega(t_1, t_2)$ is $\omega \varepsilon \Omega$ restricted to $(t_1, t_2) \cap T$.

    If $\omega, \omega' \varepsilon \Omega$ and $t_1 < t_2 < t_3$, there is an $\omega'' \varepsilon \Omega$ such that $\omega''(t_1, t_2) = \omega'(t_1, t_2)$ and $\omega''(t_2, t_3) = \omega'(t_2, t_3)$.

(b)   There is given a set A indexing a family of functions

$$F = \{f_\alpha : T \times \Omega \to Y, \alpha \varepsilon A\} \;;$$

each member of F is written explicitly as $f_\alpha(t, \omega) = y(t)$ which is the output resulting at time t from the input ω under the experiment α. Each f_α is called an input/output function and has the following properties:

(i)   (Direction of time).   There is a map $\iota : A \to T$ such that $f_\alpha(t, \omega)$ is defined for all $t > \iota(\alpha)$.

(ii) (Causality). Let $\tau, t \epsilon T$ and $\tau < t$. If $\omega, \omega' \epsilon \Omega$ and $\omega_{(\tau,t)} = \omega_{(\tau,t)}$, then $f_\alpha(t,\omega) = f_\alpha(t,\omega')$ for all α such that $\tau = _\iota(\alpha)$.

The reason why our definition (1.3.1) can be cruder than definition (1.3.2) is that we consider our system to be defined by observations of reality. Our system is not so much "an abstract summary of experimental data", as the system of definition (1.3.2) is, but rather is the data itself. We do not have to include conditions ensuring "causality" or "concatenation of inputs", because we consider these to be conditions which will be imposed on the class of models which we are willing to consider for the system, rather than conditions on the system itself. Definition (1.3.2) is undoubtedly very suitable for the deductive development of theories of system behaviour, but it is inappropriate as a starting point for the study of system identification, primarily because it assumes that the main task of system identification has already been accomplished.

To see that this is so, consider the family of input/ output functions F. Under a particular experiment α the corresponding function f_α determines the future output behaviour of the system for any acceptable input ω. But the task of system identification is precisely to determine how the system will behave in response to certain inputs. In other words, it is to determine some of the f_α's.

Furthermore, the division of observations into separate
"experiments" is properly also a part of the identification
process.   If a machine is operated, switched off, and then
operated again, the usual decision not to take account of
its behaviour when it is switched off is already a reflection
of some abstract conception we possess of the machine.   Some-
one examining a continuous record of its behaviour, and who
does not know that it has been switched off, may find a
suitable division of the record into separate "experiments"
to be far from obvious.

In our view, <u>system identification</u> is essentially the
progression from a system in the form of definition (1.3.1)
to a system in the form of definition (1.3.2).   This involves
first of all a division of the observations into experiments,
whereupon the system appears as a set of restrictions of the
functions F to certain subsets of $T \times \Omega$.   There remains the
determination, from these restrictions, of what the extensions
of the functions F are on larger subsets of $T \times \Omega$ (commonly
on all of $T \times \Omega$).   The trouble is that there are infinitely
many possibilities to choose from, even for a particular $\omega \epsilon \Omega$.

It seems as well to distinguish the problem of system
identification from that of system realisation, as defined by
Kalman, Falb and Arbib (11).   To do this, it is necessary
to quote yet another definition of "dynamical system" -
again from (11):

<u>Definition (1.3.3)</u>

A <u>dynamical system</u> (state space sense) is a composite mathematical concept defined by the following axioms:

(a)   There are given sets T, U, Ω, Y, Γ satisfying all the properties required by definition (1.3.2)

(b)   There is given a state set X and a state-transition function

$$\emptyset: \quad T \times T \times X \times \Omega \rightarrow X$$

whose value is the state x (t) $=\emptyset(t;\tau,x,\omega)\epsilon X$ resulting at time $t\epsilon T$ from the initial state $x=x(\tau)\epsilon X$ at initial time $\tau\epsilon T$ under the action of the input $\omega\epsilon\Omega$. $\emptyset$ has the following properties:

(i)   (Direction of Time). $\emptyset$ is defined for all $t\geqslant\tau$, but not necessarily for all $t<\tau$.

(ii) (Consistency). $\emptyset(t;t,x,\omega)=x$ for all $t\epsilon T$, all $x\epsilon X$, and all $\omega\epsilon\Omega$.

(iii) (Composition property).   For any $t_1<t_2<t_3$ we have

$$\emptyset(t_3;t_1,x,\omega)=\emptyset(t_3;t_2,\emptyset(t_2;t_1,x,\omega),\omega)$$

for all $x\epsilon X$ and all $\omega\epsilon\Omega$.

(iv)   (Causality).   If ω ,$\omega\epsilon\Omega$ and $\omega_{(\tau,t)}=\omega_{(\tau,t)}$, then

$$\emptyset(t;\tau,x,\omega) = \emptyset(t;\tau,x,\omega^{\hat{}}).$$

(c)   There is given a readout map $\eta:T\times X \rightarrow Y$ which defines the output $Y(t)=\eta(t,x(t))$.   The map $(\tau,t)\rightarrow Y$ given by $\sigma\rightarrow\eta(\sigma,\emptyset(\sigma;\tau,x,\omega)),\sigma\epsilon(\tau,t)$, is an output segment, that is, the restriction $\gamma_{(\tau,t)}$ of some $\gamma\epsilon\Gamma$ to (τ,t).

The problem of <u>system realisation</u> is now defined to be
the problem of constructing a dynamical system in the sense
of definition (1.3.3) from a system in the sense of definition
(1.3.2).   Kalman, Falb and Arbib (11) state that this is
"simply an abstract way of looking at the problem of scientific
model building".   We disagree.   If this were so, then .
scientific model building would be "merely" a mathematical
problem.   But the major problems with model building are
philosophical ones - questions concerning the possibility and
nature of inference, the validity of induction, in fact, all
the problems arising out of our uncertainty about the nature
of scientific method.   These problems do not arise in
connection with system realisation.   They all arise, however,
when the question of system identification, as defined above,
is considered.

That is not to say, however, that system identification
is "an abstract way of looking at scientific model building",
any more than system realisation is.   In order to be useful,
a model must not only specify the input/output functions of
a system; it must usually provide also a means of computing
them.   To do this it must take the form of a system in the
sense of definition (1.3.3).

A better abstract formulation of "scientific model
building" is, then: the problem of constructing a system in
the sense of definition (1.3.3) from a system in the sense
of definition (1.3.1).   This includes within it both the
esséntially philosophical problem of system identification,

and the essentially mathematical one of system realisation.
It is important to note that this division is a purely
conceptual one; its aim is to bring to the surface the
precise nature of the modelling problem.   It is not intended
to imply that a modeller first carries out a process of
identification, and then of realisation.   Indeed, a model
in state-space form is often used to obtain the input-output
function of the model, rather than the other way round.
But the point is that in going apparently directly from the
data to a state-space model, the modeller has implicitly
carried out a process of identification, and should be aware
of the philosophical difficulties associated with this before
accepting the model as a representation of the system.

    An illustration of the above distinctions is provided by
Ho's algorithm (11).   This algorithm constructs a system in
state-space form from a sequence of data.   The existence of
such an algorithm may appear to imply that for any sequence
of data it is possible to determine uniquely a realisation,
and that this must therefore be the "true" model of the system
generating the sequence.   However, the use of Ho's algorithm
entails the assumptions that the system input/output function
is linear, is time-invariant, and that its minimal realisation
has the smallest dimension compatible with the data sequence.
These assumptions of course specify the input/output function
completely.   Thus the decision to use Ho's algorithm itself
constitutes the process of system identification.

    The term "identification" is used somewhat differently

above than is conventional.   This is deliberate, the aim
being to distinguish clearly between the philosophical and
the mathematical problems involved in modelling.

Our definition of identification does not rely on the
precise form of definition (1.3.2).   Zadeh's concept of a
system as an "oriented abstract object" (13), - a relation
on the cartesian product of the input and output (function)
spaces - could be used in place of definition (1.3.2), as
could Windeknecht's very similar notion of a system as a
set of time functions whose attainable space (i.e. the set
of all the values of the time functions) is a relation (12).
Indeed, both of these approaches are closer in spirit to our
own definition of a system as an ordered pair of observations.
But they both share with definition (1.3.2) the essential
characteristic that makes them suitable "outputs" of the
identification process, rather than suitable "inputs":
the assumption that in any new situation the behaviour of
the system is specified.

The use of the term "system" is perhaps unfortunate,
because it can be understood in so many different senses.
In a study of modelling, definitions (1.3.2) and (1.3.3) should
really be said to define models rather than systems.   The
reason for not doing so is that we wish to reserve the term
"model" for something more specific.   On the other hand, we
persist in calling our basic data a "system", in order to
emphasise the fact that all abstract conceptions about a system
are derived from its interaction with its environment - in

other words, from our observations of it. Strictly speaking, the separation of the observations into "inputs" and "outputs" already reflects some abstract conceptions about the system, but, as has already been explained, this separation is considered to be an essential part of the "input" to the process of identification.

Such aspects as the choice of measurement scales, and even how these measurements are arranged in the arrays u_i, y_i (in definition (1.3.1)) also imply abstract preconceptions about the system. However, some preconceptions must be assumed, and we choose to regard observations as primitive. This is equivalent to considering that "objectivity" for the modeller is defined by the observations available to him.

2.   SURVEY   OF   RELATED   WORK

2.1   Complexity Measures

This survey will not attempt to cover the entire
literature on the inference of hypotheses from samples
of behaviour, since such an attempt would constitute a
literature survey of much of science and philosophy.   In
particular, the relevant philosophy of science and statistical
literature will not be included.   Instead, the survey will
trace the development of support for the thesis that the
quality of models can be associated with the shortness of
programs for computing them.   It will also examine the
work of those authors who are considered most relevant to
the present study: namely, those who have examined the nature
of scientific inference using the notion of computational
complexity.

It is convenient to begin by stating some axioms introduced by Blum (15), (16), since these provide some unifying
ideas to which most of the other work can be related.
Blum's aim is a characterisation of the complexity of
computable functions.   His method is the development of an
axiomatic theory whose theorems do not depend on the class
of computing machines which is being considered.   Our main
interest is in his axioms, which will help to classify other
work.

Let N denote the set of nonnegative integers, $\{\emptyset_i\}$
an effective listing of all partial recursive functions of
one variable, and $\{M_i\}$ a set of "machines", such that M_i

computes $\emptyset_i$. The precise nature of M_i is not specified.
It can be viewed as a program.

Axioms for Size (Blum (15)). A recursive function $| \ |$
mapping N (viewed as the set of indices) into N (viewed as
the set of sizes) is called a measure of the size of
machines, $|i|$ being called the size of M_i, if and only if
(1)   there exist at most a finite number of machines of
any given size and
(2)   there exists an effective procedure for deciding, for
any y, which machines are of size y.

Axioms for Step-Counting (Blum (16)). The set $\{\Phi_i : i = 0, 1, \ldots\}$
is a step-counting measure on $\{\emptyset_i : i = 0, 1, \ldots\}$ if and only if
(1) Φ_i is a partial recursive function
(2) $\Phi_i(n)$ converges if and only if $\emptyset_i(n)$ converges
(3) $M(i, n, m) = \begin{cases} 1 & \text{if } \Phi_i(n) = m \\ 0 & \text{otherwise} \end{cases}$ is (total) recursive.

(M is a measure on computation).

An example of a size measure is the length of a program,
if measured by the number of characters appearing in it.
The number of statements in a program is not usually a size
measure, because it violates axiom (1). The length $\ell(p)$ of
a program p is not a step-counting measure because it violates
axiom (2).

Examples of step-counting measures are the number of
statements executed by a program, and the amount of time taken

by a program (measured in CPU clock units).   These measures
are not size measures, since they depend on the argument n.

In the literature, size measures are often called
"static" complexity measures, and step-counting measures are
then called "dynamic" complexity measures.   Löfgren (17)
gives them the names "complexity of description" and
"complexity of interpretation", respectively.   In some of
the literature (e.g. Hartmanis and Hopcroft (18)), "complexity
measure" is used as a synonym for "step-counting measure".

The theorems obtained by Blum are not sharp enough to
be useful for our purposes.   This is an inevitable
consequence of the fact that they hold for any class of
machines.   For example: (Blum (15)), Let $||_M$ and $||_T$
be measures of the size of the machines M_i and T_i
respectively, where $\{M_i\}$ and $\{T_i\}$ are so ordered that M_i and
T_i both compute $\emptyset_i$.   Then there exists a recursive function
g such that, for all i:

(1) $|i|_M \leq g|i|_T$

(2) $g|i|_M \geq |i|_T$.

Thus, the sizes of M_i and T_i are not "too different".
However, the bound imposed by g is much greater than any
bound which we may find useful.   (It is only a bound in
the sense that there are functions which grow more quickly
than any recursive function).

A result of more interest to us is that there exists
a primitive recursive function whose smallest primitive

recursive derivation is considerably larger than its smallest general recursive derivation. The practical implication of this is that if we characterise modelling as a search for short algorithms, then we should be prepared to use programming languages which allow general recursion, <u>even if</u> we only wish to compute primitive recursive functions.

2.2 Algorithmic Information Theory
2.2.1 Kolmogorov Complexity

Much of the support for our view of modelling is provided by the theory of complexity and information developed by Kolmogorov (19), (20). Kolmogorov complexity is defined in terms of lengths of programs, and is therefore more akin to a static than a dynamic measure. It is, however, a property of sequences (observed behaviours in our application) rather than of machines for computing them. The results presented below have been taken from the excellent survey paper by Zvonkin and Levin (21). Proofs of most of them can be found there.

The results in which we are interested are concerned with finite binary sequences, which we call words. By means of the bijection: $\Lambda \leftrightarrow 0$, $0 \leftrightarrow 1$, $1 \leftrightarrow 2$, $00 \leftrightarrow 3$, $01 \leftrightarrow 4$,... we can regard words as nonnegative integers, and conversely (here Λ is the empty word). Thus effective procedures transforming words into words are viewed as partial recursive functions mapping integers into integers. We denote by $\ell(x)$

the length of the word x, i.e. the number of bits it contains,
and by xy the concatenation of the words x and y.   Whether
x denotes a word or the corresponding integer will always be
clear from the context.   For example, if x = OO, then by
ℓ(x) is meant ℓ(OO), but by log (x) is meant log (3).   Thus
we do not distinguish between x = OO and x = 3.   Note that
the above bijection is not the usual binary coding of
integers.   In particular, sequences which differ only in
the number of leading zeros are associated with different
integers.   Clearly, the following result holds:

$$\ell (xy) \ = \ \ell (x) + \ell (y) \qquad \ldots \ldots \ldots \ldots \ldots \ldots (2.1)$$

Also, it can easily be shown that

$$|\ell (x) - \log_2 (x)| \ \leq 1 \ , \ (x>0) \qquad \ldots \ldots \ldots \ldots \ldots (2.2)$$

Definition (2.2.1)

Let F^1 be an arbitrary partial recursive function of one
variable.   Then the Kolmogorov complexity of the word x
with respect to F^1 is defined as:

$$K_{F^1} (x) \ = \ \begin{cases} \min \ell (p) \ , \ s, \ t. \ F^1 (p) = x \\ \infty \quad \text{if no such p exists} \end{cases} \ . (2.3)$$

A generalisation of this is the concept of conditional
complexity:

Definition (2.2.2)

The conditional Kolmogorov complexity of the word x, for a

given y, with respect to the partial recursive function F^2 (of two variables) is defined to be

$$K_{F^2}(x|y) = \begin{cases} \min \ell(p) & , \text{ s.t. } F^2(p,y) = x, \\ \\ \infty & \text{if no such p exists.} \end{cases} \qquad \ldots (2.4)$$

Complexity was introduced by Kolmogorov (19), although it was also used by Solomonoff (22) and Chaitin (23), who used the terminology "the length of the shortest program required to compute x from y, using a Turing Machine which computes the function F^2". This terminology shows the intended interpretation of the definitions. p is thought of as a code or program for x, and F is thought of as a decoding device or computer. Since p is the shortest program which will do the job, the (conditional) complexity is in some sense the "smallest amount of information required to obtain x (from y), using F".

Theorem (2.2.3)

There exists a partial recursive function F_0^2 (called optimal), such that, for any partial recursive function G^2, there exists a constant C (depending only on F_0^2 and G^2), such that

$$K_{F_0^2}(x|y) \leqslant K_{G^2}(x|y) + C \ldots \ldots \ldots \ldots (2.5)$$

This theorem is due to Kolmogorov (19) and Solomonoff (22). If F_0^2 is thought of as a general purpose computer, then the theorem is easily seen to be true. C is the length of a program for F_0^2 which causes it to simulate G^2. Thus, at worst, the program for obtaining x using G^2 is prefixed by the simulation program for F_0^2, and the resulting program computes x using F_0^2. However, $K_{F_0^2}(x|y)$ may be less than $K_{G^2}(x|y)$ + C, because the greater flexibility of F_0^2 may allow x to be obtained by some shorter program. As a corollary of this theorem we have:

Theorem (2.2.4)

For any two optimal partial recursive functions F^2 and G^2, there exists a constant C (depending only on F^2 and G^2), such that

$$|K_{F^2}(x|y) - K_{G^2}(x|y)| \leqslant C \quad . . . . (2.6)$$

We henceforth assume that a fixed optimal function has been chosen, omit the subscript, and refer to Kolmogorov complexities simply as K(x) or K(x|y).

At this point it is useful to reflect on the significance of complexity, in the light of Church's Thesis. It is an intuitively appealing measure of the smallest amount of information required to obtain an entity, x, by any effective procedure. Furthermore, and remarkably, result (2.6) together with

$$\lim_{x \to \infty} K(x|y) = \infty \quad . . . . (2.7)$$

show that, for most such entities, this measure is
approximately invariant.   If the program p is considered
to be the coding of a theory which relates x to y, then
complexity gives a measure of the (intuitive) complexity
of theories, in the sense that it measures the number of
arbitrary decisions embodied in a theory.   It can therefore be viewed as measuring the quality of theories, since
it is generally believed that, other things being equal,
a simpler theory is better than a more complex one.

The following theorem has most important consequences
for modelling, as will be seen later.

Theorem (2.2.5)

The function K(x) is not partial recursive.   Moreover,
no partial recursive function $\Phi(x)$, defined on an infinite
set of points, can coincide with K(x) in the whole of its
domain of definition.

In other words, there is no effective way of obtaining
K(x) (the corresponding theorem holds also for $K(x|y)$).

Theorem (2.2.5) was first stated without proof by
Kolmogorov (19).   Zvonkin and Levin (21) prove the theorem
on the following lines.   Suppose a partial recursive function
$\phi(x)$ defined on an infinite set of points, and coincident with
K(x), does exist.   Then we can imagine a computer which
operates as follows: for each m, it uses $\Phi(x)$ to find an x
which is in the domain of definition of $\Phi(x)$, and for which

K(x)>m.   Denote this value F(m).   Then K(F(m))>m.
However, this computer (call it F) requires only to be given
m in order to find F(m).   Hence $K_F(F(m)) \leqslant \ell(m)$.   But we
know that for some C, $K(F(m)) \leqslant K_F(F(m)) + C \leqslant \ell(m) + C$.
Hence we know that, for some C, and for all m, $m < \ell(m) + C$,
which is false.   Hence such a Φ cannot exist.

The above proof relies on F(m) being general recursive.
That this can be arranged for, is shown as follows: the
domain of $\Phi(x)$ is by supposition an infinite recursively
enumerable set.   But every such set contains an infinite
recursive subset (Rogers (9), theorem 5-IV).   Hence it is
possible for F(m) to examine only integers x for which $\Phi(x)$
is defined, and therefore F(m) is defined for each m.

2.2.2 Randomness

Note that there exists a C, independent of x, such
that
$$K(x) \leqslant \ell(x) + C \quad\quad\quad\quad\quad\quad\quad\quad\quad (2.8)$$
This result says that if all else fails, we can always
compute x by making p a copy of x, together with instructions
telling the optimal computer simply to copy its input,
symbol by symbol.   This corresponds to computing x by
using a "table look-up".

Theorem (2.2.6)

The proportion of words of length $\ell(x)$ for which

$K(x) < \ell(x) - m$ does not exceed 2^{-m+1}. This means that most finite sequences have nearly maximal complexity.

Kolmogorov and Chaitin proposed that the property of maximal complexity is equivalent to the property of randomness. In other words, when we say that a sequence is "random", what we mean is that we have no way of computing it, other than by looking up its terms in a table.

This idea is a development of Church's suggestion (25), that von Mises' "Law of Excluded Gambling Systems" (26) for collectives (random sequences) can be formalised by stipulating that no effective procedure can exist for computing successful gambles on the outcomes of such sequences, and associating "effective procedure" with "partial recursive function".

Sequences are considered to be non-random if they contain sufficiently many regularities. A regularity is "any verifiable property of a sequence inherent only in a narrower class". More precisely, the measure of the set of sequences containing more than m bits of regularity cannot exceed 2^{-m}. It is essential that the regularities are verifiable, so that, as Zvonkin and Levin (21) say: "We regard as random those sequences which under any algorithmic test and in any algorithmic experiment behave as random sequences".

To explain the above paragraph more carefully, we shift our attention from finite to infinite binary sequences. We denote an infinite binary sequence by ω, and the set of all such sequences by Ω. The initial segment of ω, of length n, is denoted by $(\omega)_n$.

Definition (2.2.7)

Let P be a probability measure on Ω. A correct method
of proof of P-regularity, or P-test, is defined to be a
function F(x) which satisfies the following conditions:

(a)   It is general recursive

(b)   for m>0 , $P\{\omega : F(\omega) \geqslant m\} \leqslant 2^{-m}$,

where $F(\omega) = \sup_n F((\omega)_n)$.

$F(\omega)$, which is the "quantity of regularities" found by a
test, is taken to be the value of the test. The P-test F
is said to reject ω if $F(\omega) = \infty$.

Let Γ_x denote the set of all binary sequences whose
initial segment is the word x. A probability measure on Ω
(strictly, on the Borel σ- algebra of subsets of Ω) can
be defined by giving its values on the sets Γ_x. (To
see this, imagine infinite binary sequences to be binary
expansions of real numbers in [0,1). Then Γ_{01}, for example,
corresponds to $[\frac{1}{4},\frac{1}{2}))$.

Definition (2.2.8)

A probability measure P on Ω is computable if there
exist general recursive functions F(x,n) and G(x,n), such
that the rational number

$$\alpha_p(x,n) = \frac{F(x,n)}{G(x,n)} \quad \dots \dots \dots \dots \dots (2.9)$$

approximates $P(\Gamma_x)$ to within an accuracy of 2^{-n}.

Theorem (2.2.9)

For any computable measure $\hat{P}$ there exists a P-test F,
called <u>universal</u>, such that for any P-test G a constant C
can be found such that, for all $\omega\epsilon\Omega$,

$$G(\omega) \leq F(\omega) + C \dots \dots \dots \dots \dots (2.10)$$

Definition (2.2.10)

A sequence ω is called <u>random</u> with respect to a measure
P if it withstands any P-test.

With this definition, every sequence which is random with
respect to P, satisfies every conceivable effectively verifiable
law of probability theory, since the violation of such a law would
constitute a regularity which would be detected by some P-test.

Now consider a finte sequence x. The following construction
can be used to define the "number of regularities", $p(x)$, in x, with
respect to the uniform measure L, defined by $L\{\Gamma_x\}=2^{-\ell(x)}$. (L
corresponds to Lebesgue measure on $[0,1)$, and is the measure which
corresponds to Bernoulli sequences with generating probability $\frac{1}{2}$).

Let $F(x,n)$ denote the minimum value of the universal L-test on
words of length n beginning with x. Then

$$p(x)=\lim_{n\to\infty} F(x,n) \dots \dots \dots \dots \dots (2.11)$$

The quantity $\ell(x) - p(x)$ is analogous in several respects to complexity, and is related to it by the following theorem:

Theorem (2.2.11)

There exists a constant C, such that
$$|(\ell(x) - p(x)) - K(x)| \leqslant 4\ell(\ell(x)) + C \quad\ldots\ldots\ldots (2.12)$$
As a corollary of this we obtain, (since a random sequence has a finite number of regularities):

Theorem (2.2.12)

For any sequence ω, random with respect to L, there exists a constant C, such that
$$K((\omega)_n) \geq n - 4\ell(n) - C \quad\ldots\ldots\ldots\ldots (2.13)$$
The above development is due to Martin-Löf (24), and supports the contention that "random" is equivalent to "maximally complex".

2.2.3   Information

As was remarked in section 2.2.1, Kolmogorov complexity is an appealing measure of the intuitive concept of the "amount of information" required to obtain, or reconstruct, an object by any effective procedure.   An analogy can be discerned between complexity and entropy: it is generally accepted that entropy is the "average amount of information" required to select (i.e. predict) an event from a given probability space. Furthermore, entropy is usually taken to be a measure of the randomness of a collection of events. Section 2.2.2 suggests that Kolmogorov complexity is a suitable measure of the randomness of a sequence.

Pursuing the analogy, we make the following definition, originally proposed by Kolmogorov (19):

Definition (2.2.13)

The quantity of information in y about x is

$$I(y:x) = K(x) - K(x|y) \quad \ldots \ldots \ldots \ldots \quad (2.14)$$

This is in direct analogy with the classical Shannon information in one random variable about another, which is defined by

$$J(\eta:\xi) = H(\xi) - H(\xi|\eta) \quad \ldots \ldots \ldots \ldots \quad (2.15)$$

where H denotes entropy. This classical quantity has the following properties:

$$J(\xi:\eta) \geqslant 0 \quad \ldots \ldots \ldots \ldots \ldots \ldots \quad (2.16)$$

$$J(\xi:\xi) = H(\xi) \quad \ldots \ldots \ldots \ldots \ldots \quad (2.17)$$

$$\text{and} \quad J(\xi:\eta) = J(\eta:\xi) \quad \ldots \ldots \ldots \ldots \ldots \quad (2.18)$$

For the algorithmic quantity of information, the corresponding properties hold only approximately:

Theorem (2.2.14)

There exist positive constants, C_1, C_2, C_3, (independent of x and y) such that

$$I(x:y) \geq -C_1 \quad \ldots \ldots \ldots \ldots \ldots \ldots \ldots \quad (2.19)$$

$$|I(x:x) - K(x)| \leq C_2 \quad \ldots \ldots \ldots \ldots \ldots \ldots \quad (2.20)$$

$$|I(x:y) - I(y:x)| \leq 12\ell(K(xy)) + C_3 \quad \ldots \ldots \ldots \quad (2.21)$$

A concrete link between complexity and entropy is established by the following theorem:

Theorem (2.2.15)

Suppose that a word x, of length $\ell(x) = ir$, consists of i words, each of length r. Suppose that each of the 2^r possible words of length r (label such a word k) occurs in x with the frequency q_k (k=1, ..., 2^r). Then there exists a constant C, such that

$$K(x) \leq i(H(q_k) + \alpha(i)) + C \quad \ldots \ldots \ldots \ldots \quad (2.22)$$

$$\text{where} \quad H(q_k) = -\sum_{k=1}^{2^r} q_k \log_2 q_k \quad \ldots \ldots \ldots \ldots \quad (2.23)$$

$$\text{and} \quad \alpha(i) - C_r \frac{\ell n\, i}{i} \to 0 \text{ as } i \to \infty \quad \ldots \ldots \ldots \quad (2.24)$$

A closer connection between algorithmic and probabilistic information can be established for arbitrary ergodic stationary random processes (see Zvonkin and Levin (21)).

2.2.4   The Work of Chaitin and Schnorr

Chaitin (23) also investigated the properties of binary sequences which need maximal-length programs for their computation.   His formalism was different from that used above: he worked directly with particular Turing machines. Chaitin showed that most sequences have maximal complexity, and he obtained a result relating complexity to entropy, similar to theorem (2.2.15).   Also, he showed that the predicate $\left[\ell(x)=n \ \&K(x)<m \right]$ is not decidable.   This is a weaker result than theorem (2.2.5), since from it one can deduce that $K(x)$ is not general recursive, but not that $K(x)$ is not partial recursive.

Chaitin suggested that "maximal complexity" is an appropriate explication for "randomness".   He also hinted that complexity may be a measure of the poorness of scientific theories, arguing that "static" complexity is a more appropriate measure for this than "dynamic" complexity (he did not use these terms of course, since they were not yet current at the time).

Schnorr (27) has investigated related questions without using the concept of program.   Instead, he has examined Gödel numberings of partial recursive functions.   He has shown that there exist optimal Gödel numberings, relative to which the lowest Gödel number of each partial recursive function is rather low when compared with the lowest Gödel number of the same function relative to any other Gödel

numbering. Optimal Gödel numberings correspond closely
to Kolmogorov's optimal partial recursive functions: if
an optimal partial recursive function is universal, then
it is an optimal Gödel numbering. (However, Schnorr shows
that not all optimal functions are universal).

It should be noted that Schnorr investigates properties
of functions, whereas Kolmogorov investigates properties of
sequences, without specifying the functions which are to
compute them. Schnorr's main result is that an optimal
Gödel numbering is essentially unique in the following
sense: given any two optimal Gödel numberings, one can
obtain a Gödel number of a function f, relative to one of
the numberings, from a Gödel number of f relative to the
other numbering, by a recursive isomorphism t, such that
t and t^{-1} are linearly bounded (i.e. $\lim_{n \to \infty} \sup \frac{t(n)}{n} < \infty$ and
similarly for t^{-1}). However, Meyer (28) has shown that the
sets of <u>lowest</u> Gödel numbers relative to any two Gödel
numberings are not necessarily recursively isomorphic.
Furthermore, Meyer remarks that this result extends to sets
of "s-minimal" indices defined by <u>any</u> recursive function s
which is a size-measure in the sense of Blum.

Thus a particular interpretation of these results of
Schnorr and Meyer is the following: if one knows the
shortest program for computing some function, written in a
universal language, then one can effectively find a program,
written in some other universal language, which computes
the same function, such that the lengths of these two programs

are "not too different". But one cannot, in general,
effectively find the shortest program, written in the second
language, which computes the same function.

2.3   Asymptotic Inference.

We now turn to investigations of the process of
inference of hypotheses from observations.   It is convenient
to classify these into two categories.   In the first are
those investigations which concentrate on good "asymptotic"
results, namely on eventually obtaining good results if the
set of observations grows without limit.   The second category
consists of work which emphasises inference from a fixed,
finite sample.   In this section we examine the first of
these categories.

2.3.1   Grammatical Inference

The problem of finding a model for a behaviour is
often posed as follows: given a sample set of strings
(words) from a language, identify the grammar which generates
that language (29), (30). This formulation differs
considerably from the characterisation of the modelling
process which is developed in Chapter 3, and is mentioned
here only for completeness.   The main difference is that we
regard a model as a function from observations to observations,
whereas a grammar will in general be a relation between

observations.   In other words, a model in our sense can
be regarded as a particular set of derivations in a grammar,
rather than the grammar itself.

Another difference is that when considering grammatical
inference it is often assumed that a "teacher" is available,
who is able to supply not only those strings generated by
the grammar to be inferred, but also some strings which
cannot be generated by that grammar, and who "informs" the
inference machine which type of string it is being shown.
We do not allow this possibility, since it implies the
existence of an agent who knows the "true model" of the
system being investigated.   Gold (31) has shown that the
existence of such a teacher makes a great difference to the
asymptotic capability of a machine for inferring grammars.
A third difference is that it is usually assumed that each
string in the language will eventually be presented to such
a machine.

There are also similarities between grammatical
inference and modelling.   The most general class of grammars
investigated, the so-called "general rewriting systems" or
"type 0 grammars" are as powerful as Turing machines, in the
sense that the set of strings generated by any such
grammar is the range of some Turing machine, and conversely
(32).   Thus inferring a grammar is in general equivalent
to inferring a Turing machine.   This obviously makes the
two problems strongly related.

Another similarity is that any language can be generated
by more than one grammar, and any (finite) sample of strings
which is used for inference can belong to more than one
language.   Consequently, some criterion is needed for
choosing between rival grammars.   One way of obtaining
such a criterion is to regard the grammar as stochastic
- each production in it occurs with a certain probability.
Selection from among candidate grammars is then possible
either by using Bayes' theorem to indicate the most probable
grammar, or by using statistical testing techniques (33).
The Bayesian approach requires the specification of a priori
probabilities for the candidate grammars.

A second way of obtaining a criterion, and one which is
of more interest to us, is to choose the least complex of
the candidate grammars, (29), (34).   Another significant
difference now emerges between the problems of grammatical
inference and modelling as we understand it.   If the
complexity measure used is a "static" one, such as the sum
of the lengths of the productions in the grammar, then the
grammar which will usually be chosen is a universal grammar
which generates the language consisting of all possible
strings from the alphabet.   Consequently the complexity
measure used must include a component which is a "dynamic"
measure, such as the number of derivation steps required to
generate the sample set of strings.   When such a measure
is used, it is possible to effectively find the best grammar
(relative to the complexity measure) which generates a

particular sample (34).   In our formulation of modelling,
on the other hand, the use of a static complexity measure
is appropriate, and it is not usually possible to find the
best model in an effective manner (see chapter 3).

We shall not discuss particular algorithms for
grammatical inference, since these are not applicable to
the modelling problem posed in the next chapter.

2.3.2   Inductive Inference

We use the term "inductive inference", in contra-
distinction to "grammatical inference", to denote the problem
of inferring, from an observed behaviour, the algorithm
which produced that behaviour as output.   It will be clear
from section 2.3.1 that inductive and grammatical inference
have much in common.

In this section we review the two papers which we
consider to be by far the most significant in this field,
namely Solomonoff (22), and Blum and Blum (35).

Solomonoff considered the problem of extrapolating a
very long sequence of symbols.   He formulated this as the
problem of finding the degree of confirmation $c(a,T)$ of
the hypothesis that the sequence a will occur, given the
evidence that the sequence T has just occurred.   He considered
this degree of confirmation to be a logical probability in
the sense of Carnap (36).

Solomonoff's distinctive contribution to the solution
of this problem was that he regarded the observed and predicted

sequences as outputs of some Turing machine, and examined
the properties of those binary "programs" for this machine
which caused the observed and predicted sequences to be
computed. He was apparently the first to examine the
problem in these terms.

He presented several alternative schemes for calculating
$c(a,T)$. The first gives $c(a,T)$ a high value if the
concatenated sequence Ta can be computed by short programs
and/or if it can be computed by many programs. Short
programs are favoured because they represent simple hypotheses
about the structure of the observed sequence, while sequences
with numerous programs are favoured because of the feeling
that if they can have many alternative "causes" then they
are more "likely". The Turing machine to be used is a
"universal machine", namely one which can simulate another
universal machine by prefixing its programs with a fixed
set of "translation instructions". Such a machine
corresponds to Kolmogorov's "optimal function", and so a
theorem similar to theorem (2.2.4) holds. This is used to
show that $c(a,T)$ is fairly independent of which universal
machine is used, providing that T is long enough.

The main drawback of this scheme is that the evaluation
of $c(a,T)$ requires the summation of an infinite number of
terms, most of which are not effectively computable.
Solomonoff tries to overcome this by deriving suitable
approximations, but the necessary approximations depend
heavily on the nature of the sequences being extrapolated.

His examples are: a Bernoulli sequence, a finite-order
Markov chain, and the extrapolation of a set of strings
generated by some language (i.e. grammatical inference).

The details of the other schemes proposed are not very
relevant to us, but they all suffer from the same flaw -
they require the computation of quantities which cannot be
computed. Solomonoff's success in finding suitable
approximations for certain special cases provides evidence
in support of the conceptual validity of his schemes, but
it does not make them into practical "inference machines".
To make this point clearer: the amount of knowledge
needed, in order to make a suitable approximation, is
sufficient for proceeding direct to a solution, without any
need to use Solomonoff's procedures.

It is important to note that $c(a,T)$ is not based on
just a single "best" program, but on the set of all
programs that compute the sequence Ta. Thus Solomonff's
prediction is based on a "weighting" of all possible
models, rather than on the "best" model.

The problem investigated by Blum and Blum (35) is
the same as that investigated by Solomonoff, but the formal
setting is slightly different. The observations are assumed
to be a set of pairs (x,y), and the law which explains them
to be an algorithm for computing $f(x)=y$. The problem is to
find an algorithm that computes f. This formulation of
inductive inference is very close to the problem of modelling,
as formulated in Chapters 1 and 3.

Blum and Blum attempt to characterise those functions
f which it is possible to identify - that is, those functions
for which it is possible to infer a correct algorithm.
The characterisations which they obtain are in terms of step-
counting (dynamic) complexity measures.   Generally speaking,
it is possible to infer a function if it is not too difficult
to compute it.   These results give the clearest connections
between inference and complexity that have been established
to date.   However, a step-counting measure is very
different in nature from a size measure, so these results
can not be used to support the hypothesis that small models
are good models.

Nevertheless, the authors state their conviction that
"to have quality, a hypothesis n bits long must explain more
than could merely be encoded . . . in that many bits".
The machines which they construct in their proofs invariably
identify "small" algorithms, in the sense that they search
through a Gödel numbering of all partial recursive functions
by a process of enumeration - the Gödel number of an
algorithm is of course a size measure for it.   (It is not
possible for them to find the smallest suitable Gödel number,
in general, but the search nevertheless occurs in order of
increasing size).

In one example, Blum and Blum employ an interesting
construction to ensure that the inferred algorithm is small
enough to be meaningful.   A machine is to identify some
arbitrarily difficult 0-1 valued recursive functions. (A machine

converges in the limit to i if it eventually outputs i and
then never outputs a different number, upon being presented
with some infinite sequence of pairs.   It can identify f
if, whenever it is given a complete sequence of pairs
$(x, f(x))$, it converges to i and the partial recursive function
$\emptyset_i$ is an extension of f).   The machine works as follows.
If its last conjecture is i, and it finds that $\emptyset_i(y)$ is
defined and $\emptyset_i(y) \neq f(y)$, then it conjectures i+1.   On the
other hand, if $\emptyset_i(y) = f(y)$ for all y<x then it tests $\emptyset_i(x)$ in
the following manner.   First it constructs an upper bound,
then it tests whether $\emptyset_i(x) = f(x)$ within this upper bound.

If so, it accepts i, otherwise it conjectures i+1.   The
interesting part is the construction of the upper bound:
first a recursive function h is fixed.   Then a j is searched
for, such that $\emptyset_j$ converges to f on inputs $0, 1, \ldots, \max(2j, 2x)$
faster (i.e. in fewer steps) than $\emptyset_i$ converges on x.   Let
$\{\Phi_i\}$ be the set of step-counting measures being used.   When
a suitable j is found, take the upper bound to be max
$(h(x, f(x)), \max(\Phi_j(y): y \leq \max(2j, 2x)))$.

   Thus conjecture i is abandoned if it is discovered
that some algorithm computes a restriction of the function
to be inferred more quickly than i, for a considerably
larger set of values.   Now the reason why j can be regarded
as a potentially meaningful explanation of the data is
this: think of j as the jth program for some universal

machine.   If it is written in a binary alphabet its
length is roughly $\log_2 j$, and it is therefore not large
enough to store $(f(0),f(1),...,f(2j))$ in a look-up
table (recall that $f(n)\epsilon\{0,1\}$).

2.4   Small-Sample Inference
2.4.1 Wrinch and Jeffreys

In 1921 Wrinch and Jeffreys (37) proposed that competing
models of a set of observations should be assessed on the
basis of simplicity.   They suggested that any model in
physics could be formulated as a differential equation;
if two differential equations explained the same set of data,
then the one with the fewer parameters should be preferred.
In fact they proposed assigning probabilities to models in
this form, the probability being a decreasing function of
the number of parameters.   Popper (38) suggested a similar
but vaguer scheme, but argued that Wrinch and Jeffreys
should have regarded the simpler models as the **less** probable
ones.   This argument seems to arise from an almost wilful
confusion of two separate concepts, and it is convenient to
clarify these at this stage.

It seems clear that Wrinch and Jeffreys use the term
"probability" in the sense of "degree of confirmation".
That is, having observed some data, and having conjectured
some hypotheses which are capable of explaining the data,
they are assessing the relative likelihoods of the various

predictions entailed by those hypotheses. This is equivalent
to assessing the quality of the competing hypotheses. In
this sense "probability" is synonymous with Popper's
"degree of corroboration", which, he insists, increases with
simplicity. Thus Popper's view is consistent with that of
Wrinch and Jeffreys, if this interpretation of "probability"
is admitted. (Popper argues, however, that"degree of
corroboration" cannot be interpreted as a probability. This
is probably correct, but it is a separate point. An
argument which supports it is given in Chapter 8).

Whereas Wrinch and Jeffreys refer to something like
"the probability that this model correctly predicts future
behaviour, in the light of the behaviour already observed
and the models which have been conjectured", Popper uses
"probability" to mean, roughly, "given that some observations
are to be made, what is the probability that it will be
possible to explain them using a model with n parameters?"
Now it is quite reasonable, intuitively, that this probability
should increase with the number of parameters, as Popper
suggests (if it is defined at all).

Thus no conflict arises between Wrinch and Jeffreys
and Popper, providing that it is borne in mind that Wrinch
and Jeffreys use "probability" to indicate a property of a
model, in the light of observations, whereas Popper uses it
to denote an a priori property of the observations. This
thesis is concerned entirely with the properties of models.

As it stands, the criterion of quality proposed by
Wrinch and Jeffreys is not a practical one. Empirical
data would usually need a very high order differential
equation model to fit it. In practice, some approximation
is tolerated in order to allow a simpler model, and some
trade-off between approximation and complexity has to be
made. Also, the only support for their proposal is the
intuitive feeling that the number of parameters is the
appropriate measure of complexity. Apart from this, however,
the criterion of Wrinch and Jeffreys is very similar to the
proposed criterion of model quality which is introduced
in Chapter 3, and is close to being a special case of it.

2.4.2   Gaines

Gaines has recently proposed a formulation of the
general system identification problem (14). The characterisation of Chapter 3 can be viewed as a special case of
Gaines' proposals. Gaines considers an identification
problem to be defined by an observed behaviour, a class
of models which are of interest, an arbitrary but fixed
partial ordering of models in this class (this ordering
being called "complexity"), and an arbitrary partial ordering
of models in this class which is induced by the particular
behaviour being observed (this ordering being called
"approximation"). The "admissible subset" of models is
then the set of models which has the property that, if m is

a member of this admissible subset, then no model exists
which is both less complex than m, and gives a better
approximation to the behaviour than m.   Gaines considers
this admissible subset to be the solution to the identification
problem.   Which model is to be preferred from among those
in the admissible subset depends on the particular circumstances
of each modelling exercise.   We go further than this in our
characterisation, and in effect propose a particular trade-
off between approximation and complexity.

A point which will be of interest later (in Chapter 8)
is that Gaines associates changes in the complexity and
approximation ordering relations with Kuhn's "scientific
revolutions" (39).   Thus a change in either or both of
these relations corresponds to a new "view of science".

Gaines has investigated the special case of the
behaviours to be modelled being finite sequences of arbitrary
symbols, the class of models being finite state automata
(both deterministic and probabilistic),   the measure of
complexity being the number of states, and several measures
of approximation based on the Hamming distance between the
observed and computed behaviours.   In this case the set of
admissible models is effectively computable.   However, it
is not practical to identify most behaviours which are of
interest to us in this manner, partly because of excessive
computational requirements, but mainly because many behaviours,
which can be produced by compact algorithms, can be modelled
only by maximally complex finite-state automata.   Thus the

use of this class of models does not lead to a discernible
structure when used for the identification of many simple
behaviours.   An example of such a behaviour is that produced
by the program:

```
     n:= 1;
loop:n:= n*(n+1);
     write  (n);
     goto loop;
```

namely,the sequence 2,6,42,1806,.....   This program cannot
be correctly implemented by either a deterministic or stochastic
finite state automaton.

2.4.3 Löfgren

Löfgren (40) has made the rather unlikely suggestion
that"as soon as a scientist believes that he has produced a
theory for some phenomenon, he should try to formalise the
theory so as to make it effectively communicable".   By
"formalise" Löfgren means that the theory should be translated
into one of the formal logical systems!   However, if
"formalise" is reinterpreted to mean that the theory should
be expressed as an algorithm for computing the observed
phenomenon, then Löfgren's views on the quality of theories
become virtually identical with our views on models.

For example, his key hypothesis is: Let S and S' be
two formal theories with the same logical basis, both of
which explain one set of experimental facts with comparatively

short and few proper axioms and proper rules of inference.
Then, if both S and S' predict further experimental results
(as being proper theorems or not), the theory with the
simplest proper axioms and proper rules has the greater
predictive power, in the sense that its predictions are
more likely to agree with further experiments.   The simplicity
of the proper axioms and proper rules of inference can be
measured by the total length of all corresponding well-
formed formulae".   This hypothesis will be seen to be very
close to the basic idea underlying this thesis, that the
quality of a model can be measured by the shortness of the
program required to realise it.

Also, Löfgren has anticipated one of the key concepts
which will be introduced later: "Let $\{h_1,...,h_n\}$ be a set
of experimentally obtained facts.   This set can be considered
the theoremhood of a formal theory S_o with $\{h_1,...,h_n\}$ as the
set of axioms, with no rules of inference, and thus with a
logical basis which is empty .... Such a mere listing of
the experimentally obtained facts has no predictive power
whatsoever".   The theory S_o corresponds to what we shall
later call the "trivial model", and which will serve as a
standard against which the quality of models will be measured.

Löfgren also investigated the connection between
randomness of a sequence and the length of program required
to compute it.   He proved a weaker form of theorem (2.2.5)
by using the recursion theorem (see Rogers (9)), and showed

that there can be no algorithm for finding a sequence
whose Kolmogorov complexity is greater than a given value.

2.5   Inference of Parameters of Gauss-Markov Process

Recently, Rissanen (41) has investigated the estimation
of the structure, as well as the real-valued parameters, of
a Gauss-Markov process, by using the idea that a good model
is a short description of the observed data.

Suppose that the set $y=(y(0),\ldots,y(N-1))^T$ has been
observed, and that it was generated by the process

$$x(t+1) = Ax(t)+Be(t)$$
$$y(t) = Cx(t)+e(t), \quad x(0)=0,$$
$$t=0,\ldots,N-1.$$

Then $((A,B,C); \quad e(o),\ldots,e(N-1))$ can be considered to be a
description of y, since y can be recovered from it by
using y=Te, where

$$T= \begin{bmatrix} I & & \\ CB & I & \\ CAB & CB & I \\ & \vdots & \end{bmatrix}$$

y is supposed to be recorded with an accuracy of a fixed
number of fractional bits.   It is therefore meaningful to
consider the shortest description of y.   It is known that
the expected length of such a description is bounded below
by the entropy of y.   Rissanen demonstrates that, if the
triple (A,B,C) is such that the sequence e is serially

uncorrelated, and if the distribution of the observations is Gaussian, then e can be encoded as a binary string whose length nearly attains the lower bound.   He also shows that in the scalar case the bound is nearly reached by simply writing each e(t) as a binary number.

There are many triples (A,B,C) which give rise to the same sequence e.   Each triple (A,B,C) can be identified with a pair (s,θ), where s is an integer, and θ is a vector of real-valued parameters (42).   s represents the structure of (A,B,C), and is in fact an enumeration of a set of integers, which define the order of the system and the positions of the components of θ.   Thus another form of description of y is a triple(s,θ,ê), where ê is a code of e. If (s*,θ*) make e uncorrelated, then a description

   z* = (s*,θ*,ê*)

is called a <u>concise</u> description of y, where ê* is the shortest coding of e.

An <u>estimator</u> y → (s,θ) is <u>concise</u> if the length of the associated description $(\hat{s},\hat{\theta},\hat{e})$ is minimised, where $\hat{s},\hat{\theta}$ are the best codes for s,θ.   Thus, in addition to minimising the length of ê, a concise estimator also minimises the mean length of $(\hat{s},\hat{\theta})$.   Given a structure s, a candidate asymptotically concise estimator is one which at least minimises the estimated entropy of e.   For a Gaussian distribution, such an estimator coincides with the maximum likelihood estimator.

If s* is the true structure of the process, then the estimates of θ given by this estimator are asymptotically

Gaussian at s=s*. Consequently it is possible to estimate the entropy of θ, conditional on s*. Furthermore, $H(\theta|s) \geqslant H(\theta|s*)$, where H denotes entropy, and Rissanen shows how $H(\theta|s)$ may be estimated. If $H(\theta|s)$ is minimised, then not only has a θ been found whose shortest expected length of description is the shortest possible, but the associated s must also be the true structure.

Thus an estimator which minimises an estimate of $H(e) + \frac{1}{N}H(\theta|s)$, where the minimisation is over (s,θ), is an asymptotically concise estimator, namely one which eventually gives the shortest possible description of the observed data y, and which also estimates correctly the structure s. Furthermore, Rissanen shows that $H(\theta|s)$ is small when the error covariance is highly sensitive to variations in θ.

2.6   Summary

The above survey has, hopefully, filled in the background to the treatment of models which will be developed in this thesis - that is, a treatment based on ideas of computational complexity. Blum's axioms help to distinguish two very different notions of complexity - the complexity required to describe something, and the complexity of interpreting that description.

A detailed review has been given of the theory developed from the notion of Kolmogorov complexity. This complexity is shown to be not effectively computable. The results

which have been given support the idea that complexity
represents the amount of information required to obtain
an   entity by any effective procedure.   They also
indicate that the property of being maximally complex is
equivalent to the property of being random.   These ideas
are the chief support of our approach to modelling.
Kolmogorov complexity is asymptotically invariant with
respect to a wide class of universal computers, but it will
be seen later that this invariance property does not appear
sufficiently quickly for the purposes of practical model
assessment, and the choice of computer (in fact, of language)
constitutes a major problem.

Aspects of grammatical inference have been briefly
examined.   Although grammatical inference is conceptually
very close to the inference of any type of model from
observations, the statement of the problem is sufficiently
different from what we understand as the modelling problem,
to make the nature of possible solutions very different.
The use of dynamic complexity measures is appropriate (in
fact, essential) for grammatical inference, and this allows
an effective procedure to exist for finding the "best"
grammar, within a sufficiently restricted class.   (Admittedly
the procedures attempted so far are based on exhaustive
enumeration, and so are not practical, but the problem can
at least be solved in principle).

Solomonoff's approach to inductive inference is the

progenitor of the work reported in this thesis.   However,
Solomonoff emphasised a quantitative measure of "degree of
confirmation" of a theory, and this required the summation
of uncomputable terms.   A major difference between Solomonoff's
work and ours is that Solomonoff considered that "degree of
confirmation" is decided by all the possible programs for a
behaviour, whereas we consider it to be determined by only
one - the best one available.

The paper of Blum and Blum has been mentioned because
it is clearly most important for inductive inference
generally. However, it is not very relevant to this thesis
because all its results refer to the asymptotic situation,
whereas we are concerned with the situation of having a fixed,
finite sample of behaviour.

Under the heading "Small-Sample Inference" we considered
three papers.   That of Wrinch and Jeffreys is interesting
because it can almost be regarded as a special case of our
proposals.   Gaines suggests that any system identification
problem has the same basic features, and these features can
be discerned in our formulation of the modelling problem.
Finally, Löfgren suggests a scheme for assessing theories
which is very similar to our scheme for assessing models.

The approaches to inference which we have surveyed have
a characteristic in common: while each of them seems to
provide a more or less plausible account of the abstract nature
of inference, none of them can be applied to a useful

range of practical models.   In Chapter 3 we shall develop
a view of modelling which is based on ideas of complexity,
and which can be directly applied to models of the type
encountered in control studies.

In section 2.5 we have examined the work of Rissanen,
which implicitly assumes a very similar view of modelling.
By examining a particular probabilistically described
situation, Rissanen can replace explicit consideration of
complexity by an investigation of the classical Shannon
entropy associated with the model.   For the case of a
Gauss-Markov process, the search for the shortest model of
the data leads to an extension of maximum-likelihood
estimation.   This work gives strong support to the plausibility
of the characterisation which we shall develop below.

3. A CHARACTERISATION OF MODELLING.

3.1 Introduction

In this chapter we propose a partial solution to the modelling problem introduced in section 1.3. The solution is partial because it does not provide an algorithm for obtaining a model for a system, but only a criterion for choosing between competing models. However, a consequence of our characterisation of modelling is that such an algorithm cannot exist. Our solution is, therefore, as complete as can be expected. (More complete solutions can be obtained if the class of candidate models is sufficiently restricted).

3.2 Systems

Systems have already been introduced in definition (1.3.1). For each system S there exists a smallest positive integer n, such that multiplying every u_j^i, y_j^i appearing in S by n (except for blanks, which are left unchanged) results in an <u>integer system</u> $S_I = (U_I, Y_I)$; that is, a system as defined previously, but now each observation is either an integer or a blank. Each integer system S_I can be identified with a countable equivalence class of systems, each of which is mapped into S_I by the above transformation. We shall consider the integer systems to be the entities which are to be modelled. The prefix "integer" will usually be omitted.

We need to establish that the set of integer systems can be identified with the set of nonnegative integers. However, we do not need to know the details of the correspondence.

Theorem (3.2.1)

There exists an effective bijection between the set of integer systems and the set of nonnegative integers.

<u>Proof.</u>   For each u_j^i in an integer system S_I, write 0 if $u_j^i = b$, $u_j^i + 1$ if $u_j^i \geqslant 0$, and u_j^i otherwise.   Similarly for each y_j^i. S_I has now been replaced by an ordered pair of arrays of integers.   Each integer can be mapped into a nonnegative integer by the function p, where:

$$p(n) = 2|n| \text{ if } n \leq 0,$$
$$2n-1 \text{ if } n > 0.$$

Instead of considering S_I, we can now consider S_I^+, where

$$S_I^+ = (((u_1^1, \ldots, u_{\ell_1}^1), \ldots, (u_1^M, \ldots, u_{\ell_M}^M)), ((y_1^1, \ldots, y_{m_1}^1), \ldots, (y_1^N, \ldots, y_{m_N}^N))),$$

and where each u_j^i, y_j^i is now a nonnegative integer.

Rogers (9) demonstrates the existence of a recursive bijection $\tau: \mathbb{N} \times \mathbb{N} \to \mathbb{N}$, which he calls a <u>pairing function</u>, and he uses it to define recursively a coding of k-tuples of nonnegative integers onto the nonnegative integers:

$$\tau^k(n_1, n_2, \ldots, n_k) = \tau(n_1, \tau^{k-1}(n_2, \ldots, n_k)),$$

with $\tau^1(n) = n$.

If S_I^+ is now replaced by an "indexed" version:

$$((M,(\ell_1 u_1^1,\ldots,u_{\ell_1}^1),\ldots,(\ell_M,u_1^M,\ldots,u_{\ell_M}^M)),(N,(m_1,y_1^1,\ldots,y_{m_1}^1),\ldots,$$
$$(m_N,y_1^N,\ldots,y_{m_N}^N)))$$

then it can be coded as the single nonnegative integer:

$$\tau(\tau^{M+1}(M,\tau^{\ell+1}(\ell_1,\ldots,u_{\ell_1}),\ldots,\tau^{\ell_M+1}(\ell_M,\ldots,u_{\ell_M}^M)),\tau^{N+1}(N,\tau^{m_1+1}$$
$$(m_1,\ldots,y_{M_1}^1),\ldots,\tau^{m_N+1}(m_N,\ldots,y_{m_N}^N))).$$

Since an "indexed" version of S_I^+ has been coded, it is

possible to recover S_I^+ uniquely from this single integer

(since τ is invertible). Also, it is obviously possible

to recover S_I from S_I^+. It is now possible to search through

the sequence $(0,1,2,\ldots)$ for the sequence of integers

$(n_0,n_1,n_2,\ldots)$ from which a valid S_I^+ can be recovered. By

establishing the correspondence $n_0 \leftrightarrow 0$, $n_1 \leftrightarrow 1,\ldots$, the required

bijection is obtained.

Theorem (3.2.1) establishes that a nonnegative integer

can be uniquely associated with every integer system, and

that it therefore makes sense to refer to "the ith (integer)

system, S_i". But it was pointed out in section 1.3 that

every input observation set U (except U=b) and every output

observation set can itself be regarded as a system.

Consequently a nonnegative integer can be associated with

each of these sets. As in section 2.2, we shall frequently

not distinguish between a system and the integer associated

with it.

3.3 Models

It is tempting to regard a model as an algorithm
which operates on the input observation set to produce the
output observation set of a system.   This would be sufficient
for many purposes, in particular for the treatment of
deterministic models.   However, in some cases it is
desirable to allow models to use some of the output
observations for computing others.   The most common example
of this occurs in the use of models for the prediction of
system behaviour in a stochastic environment.   In this case
it would be unreasonable not to allow the model the use of
the most recent information on system behaviour.   (In the
deterministic case this is unnecessary, since the model can
determine system behaviour from the initial conditions and
the input history).

It is therefore necessary to allow a model to be an
effective procedure which operates on some outputs as well
as on inputs, in order to compute the output.   This capability must be restricted, however, if one is to exclude,
for instance, models which compute the output observation
set simply by copying it, or those which use future observations
to compute previous ones.   We accomplish this restriction in
a rather roundabout way.   Instead of defining models so as
to exclude these useless types of algorithms, we define models
with reference to certain sets of subsets of the observations.
These sets determine which input and output observations may
be used for the computation of each output.   Restriction to

the classes of models which are of interest is then
accomplished by suitably delimiting these sets.

We distinguish between abstract models, which are
partial recursive functions, and concrete models, which are
computer programs.   This terminology follows Chaitin (43).

Definition (3.3.1)

Given a system $S=(U,Y)$, let $A=\{A_i\}$ be a set of ordered
subsets of U, let $B=\{B_i\}$ be a set of ordered subsets of Y,
and let $C=\{C_i\}$ be a set of m disjoint ordered subsets of Y,
which is complete in the sense that every $y_j \in Y$ occurs
in some C_i.   The ordering of the elements of A_i, B_i and
C_i is to be the same as their ordering in U and Y.   Let D
be a set of ordered pairs $D_i=(A_j, B_k)$, $(i=1, \ldots, m)$, and
finally let E be a set of ordered pairs $E_i=(D_i, C_i)$, $(i=1, \ldots, m)$.

Then an _abstract E-model_ of the system $S=(U,Y)$ is a
partial recursive function $M: \mathbb{N} \times \mathbb{N} \to \mathbb{N}$, such that, for each
$E_i \in E, (i=1, \ldots, m)$,
$$M(i, D_i) = C_i \quad \ldots \ldots \ldots \ldots \ldots \ldots \quad (3.1)$$

This definition is best illuminated by some examples.
We use the notation $S = ((u_1, \ldots, u_N), (y_1, \ldots, y_N))$.

Example (3.3.2)

If only non-dynamic models were of interest, we might

specify the sets A,B,C,D,E as follows:

$$A_i = u_i \quad , \quad i=1, \ldots, N,$$

$$B_i = \emptyset \quad , \quad i=1, \quad (\emptyset \text{ denotes the empty set})$$

$$C_i = y_i \quad , i=1, \ldots, N,$$

$$D_i = (A_i, B_1) = (u_i, \emptyset) \quad , i=1, \ldots, N,$$

$$E_i = (D_i, C_i) = (u_i, y_i), \quad i=1, \ldots, N,$$

so that $E = \{(u_i, y_i) : i=1, \ldots, N\}$.

In this case, an abstract E-model of S is a partial recursive function M, such that

$$M(i, (u_i, \emptyset)) = y_i \quad \text{for } i=1, \ldots, N.$$

Example (3.3.3)

If we were interested in dynamical, deterministic, finite -memory (say, two-period) models, a suitable specification of the sets might be:

$$A = \{\emptyset, (u_1, u_2, u_3), (u_2, u_3, u_4), \ldots, (u_{N-2}, u_{N-1}, u_N)\},$$

$$B = \{\emptyset\}$$

$$C_i = y_i \quad , i=1, \ldots, N,$$

$$D_i = \begin{cases} (\emptyset, \emptyset) & \text{for } i=1,2, \\ ((u_{i-2}, u_{i-1}, u_i), \emptyset) & \text{for } i=3, \ldots, N \end{cases}$$

$$E_i = \begin{cases} ((\emptyset, \emptyset), y_i) & \text{for } i=1,2, \\ (((u_{i-2}, u_{i-1}, u_i), \emptyset), y_i) & \text{for } i=3, \ldots, N \end{cases}$$

This time an abstract E-model of S is a partial recursive function M, such that

$$M(i,(\emptyset,\emptyset)) = y_i \quad , \quad \text{for } i=1,2,$$

and $M(i, ((u_{i-2},u_{i-1},u_i),\emptyset)) = y_i$ for $i=3, \ldots, N$.

Example (3.3.4)

If we were interested in one-step-ahead predicting models, which were allowed to use all past observations, we could specify the sets as follows: (We take U = b for simplicity in this case)

$$A = \{\emptyset\}$$

$$B = \{\emptyset, y_1, (y_1,y_2), (y_1,y_2,y_3),\ldots,(y_1,\ldots,y_{N-1})\}$$

$$C_i = y_i \quad , \quad i=1,\ldots,N,$$

$$D_i = \begin{cases} (\emptyset,\emptyset) & \text{for } i=1, \\ (\emptyset,(y_1,\ldots,y_{i-1}) & \text{for } i=2,\ldots,N, \end{cases}$$

$$E_i = \begin{cases} ((\emptyset,\emptyset),y_i) & \text{for } i=1, \\ ((\emptyset,(y_1,\ldots,y_{i-1}),y_i) & \text{for } i=2,\ldots,N. \end{cases}$$

In this case, an abstract E-model of S is a partial recursive function M, such that

$$M(1,(\emptyset,\emptyset)) = y_1$$

and $M(i,(\emptyset,(y_1,\ldots,y_{i-1}))) = y_i$ for $i=2,\ldots,N$.

M takes two arguments so that the first can act as an index, which "tells" M which block of observations it is operating on. The significance of this can be seen most clearly if we anticipate a little, and consider M to be a

computer program. The program may be designed to operate
on different blocks in different ways, for example by operating
on them with different subroutines. In order to do this,
the program must be told which block it is currently
operating on. It would be possible to allow M only one
argument, but it would then be necessary to regard a model
as a set of such partial recursive functions, each of which
corresponded to a different computational algorithm.

It is not usual to specify the class of models which is
of interest by specifying the sets A,B,C,D,E in the manner
of the above examples, primarily because of the risk of
inadvertently excluding some models which may be of interest.
A more usual method is to specify conditions which these sets
must satisfy. For example, if it were desired to exclude
the possibility of a model merely "copying" the output
observation set, it would be sufficient to consider only
those E-models for which E satisfied the condition:

for each $E_i \varepsilon E$, $(y_k \varepsilon B_j \& B_j \varepsilon D_i) \Rightarrow y_k \notin C_i$.

Here "$y_k \varepsilon B_j$" denotes "$B_j = (\ldots, y_k, \ldots)$," and "$B_j \varepsilon D_i$" denotes
"$D_i = (\ldots, B_j)$".

Similarly, if models which use future observations to
compute previous ones are of no interest, they can be
excluded by imposing suitable conditions on E:

Definition (3.3.5)

A set E, defined as in definition (3.3.1), is nonanticipative if, for each $E_i \in E$:

    (i) $\max(p:y_p \in B_j \& B_j \in D_i) < \min(p:y_p \in C_i)$

and   (ii) $\max(p:u_p \in A_j \& A_j \in D_i) \leqslant \min(p:y_p \in C_i)$.

An E-model will be said to be nonanticipative if E is nonanticipative.

We wish to consider models to be computer programs. We formalise a computer (together with a programming language) as a (not necessarily universal) 3-place partial recursive function F, and a program as the first argument of this function.

Definition (3.3.6)

A <u>concrete (F,E)- model</u> of the system S=(U,Y) is an integer p, such that

$$F(p,i,D_i) = C_i \quad . \; . \; . \; . \; . \; . \; . \; . \; . \; . \; . \; . \; . \; . \; . \; . \; . \quad (3.2)$$

for every $E_i \in E$, where C_i, D_i, E_i, E are defined as in definition (3.3.1), and $F:\mathbb{N} \times \mathbb{N} \times \mathbb{N} \to \mathbb{N}$ is a 3-place partial recursive function.

Theorem (3.3.7)

There exists an F, such that to every abstract E-model M of S there corresponds a concrete (F,E) -model p of S, such that

$$F(p,x,y) = M(x,y) \quad . \; . \; . \; . \; . \; . \; . \; . \; . \; . \; . \; . \; . \quad (3.3)$$

for all $x,y \in \mathbb{N}$.

Proof: Choose F to be any 3-place universal partial
recursive function.   Then there exists p, such that
$F(p,x,y)=M(x,y)$ for all $x,y \epsilon \mathbb{N}$ (Rogers(9),p.22).   But
$M(i,D_i)=C_i$ for every $E_i \epsilon E$.   Therefore $F(p,i,D_i)=C_i$ for
every $E_i \epsilon E$.   Consequently p is an (F,E)-model of S.

Theorem (3.3.8)

To every concrete (F,E) -model p of S there corresponds
an   abstract E-model M of S, such that

$$M(x,y)=F(p,x,y) \quad \ldots \ldots \ldots \ldots \ldots \ldots \quad (3.4)$$

for all $x,y \epsilon \mathbb{N}$.

Proof: By the s-m-n theorem (Rogers(9),p.23), there exists
a 2-place partial recursive function M, such that
$M(x,y)=F(p,x,y)$ for all $x,y \epsilon \mathbb{N}$.   But $F(p,i,D_i)=C_i$ for

every $E_i \epsilon E$.   Consequently $M(i,D_i)=C_i$ for every $E_i \epsilon E$, and

therefore M is an abstract E -model of S.

Theorems(3.3.7) and (3.3.8) show that abstract and
concrete models are essentially equivalent.   The function
F which appears in definition (3.3.6) can be associated with
some computing facility, by invoking Church's Thesis.
The nonnegative integers can be used to enumerate the programs
for such a facility in some standard manner (cf section 2.2.1).
Thus concrete models can be associated with programs.

Note that every system has infinitely many abstract
E-models, and that to each of these abstract E-models there
will usually correspond infinitely many concrete (F,E)

-models (for a fixed F). If F is universal then there will always correspond infinitely many concrete (F,E) -models to each abstract E -model, since the enumeration of the set of programs for F then constitutes a Gödel numbering for the partial recursive functions: it is known that in every Gödel numbering, each partial recursive function has infintely many Gödel numbers. The task of modelling is to find an appropriate abstract model for a system. In practice this requires the postulation and examination of concrete models (programs) for it.

According to the discussion of section 1.3, the goal of the modelling exercise is a dynamical system in the sense of definition (1.3.3). To reconcile this with the above statement, we point out that a nonanticipative abstract E-model M can be regarded as such a dynamical system , providing that each C_i has the form $C_i = (y_k, y_{k+1}, \ldots, y_{k+n_i})$

(in other words, there are no gaps in the set C_i). It suffices to take the time set T to be the set of integers $(0,1,2,\ldots,m)$ with the usual ordering, to take the initial time $\tau = 1$, and to identify time t with the computation of C_t (using the terminology of definitions (1.3.3) and (3.3.1)). The state at time t can be taken to be the input and output history $x(t) = ((u_1, u_2, \ldots, u_j), (y_1, y_2, \ldots, y_k))$, where $j = \max(p: u_p \varepsilon A_q \& A_q \varepsilon D_i, i \leqslant t)$ and $k = \max(p: y_p \varepsilon B_q \& B_q \varepsilon D_i, i \leqslant t)$. The input at time t is the sequence $\omega(t) = (u_{j+1}, \ldots, u_r)$ where $r = \max(p: u_p \varepsilon A_q \& A_q \varepsilon D_t)$, and the output is C_t. Since E is nonanticipative, all the elements of D_{t+1} appear in

$(x(t), \omega(t+1))$, and so D_{t+1} can be "assembled" from

$(x(t), \omega(t+1))$.   Let Ψ be an algorithm for doing this.

Then Ψ and M together determine the state transition function $\emptyset$:

$$\emptyset(t+1;t,((u_1,\ldots,u_j),(y_1,\ldots,y_k)),\omega(t+1)) =$$

$$((u_1,\ldots,u_j,\omega(t+1)),(y_1,\ldots,y_k,M(t+1,\Psi(t+1,x(t),\omega(t+1))))).$$

The initial state $x(0)$ is a pair of empty sequences, and

the readout map η is defined by $\eta(t,((u_1,\ldots,u_\ell),(y_1,\ldots,y_m,c_t))) =$

$$c_t, \text{ for } t > 0.$$

Conventionally, models are allowed to approximate system behaviour, rather than reproduce it exactly.   Definitions (3.3.1) and (3.3.6) however, require models to compute the observed system behaviour exactly.   This does not mean that the class of models which we can treat is any smaller than the class of models which are usually of interest.   It merely means that whereas a conventional model may reproduce a system behaviour approximately, the corresponding model in our formalism has the additional task of generating the "corrections" which must be applied to the approximate behaviour, in order to produce the exact system behaviour.

Fig. 2 shows the correspondence between a type of conventional model commonly encountered in control studies, and a model which satisfies definitions (3.3.1) and (3.3.6).

It will be recalled that theorem (2.2.12) suggests that "random" is equivalent to "can be computed only by using a table look-up".   If a model is considered to be a summary

of knowledge about a system, then those computations of the
model which have to be performed by using a table look-up
correspond to those aspects of the system behaviour which
are not understood, and cannot be predicted - in fact,
those that appear to be random.   The role of these computations
may be very different from that shown in fig. 2.   For example,
if they are "corrections", they need not be additive.   But
more generally, the terms computed by table look-up need not
play the role of "corrections".   They may, for instance,
be parameters, which would conventionally be viewed as
"randomly varying".

3.4 Criterion of Quality

The third component of our characterisation of modelling
is a criterion of quality of a model.

Let F represent a computing facility, together with a
programming language.   Let c be an injective function from
the integers to the set of strings of terminals in the programming language, which is used to represent the integers in
programs.   (c therefore is included in the definition of the
programming language F.   The definition of programming
languages is reviewed in Appendix A;   more details of c are
given in Chapter 7).   Let S be an integer system as defined
in section 3.2, with input and output observations u_j^i, y_j^i.

Definition (3.4.1)

The trivial F model of S is the shortest program which

is a concrete (F,E)-model of S, such that each $c(y^i_j)$

appears in it, where the minimisation of length ranges over

all possible sets E (defined by def. (3.3.1)).

It is assumed that the length of a program is measured
by the number of terminals appearing in it.

The trivial model of a system is one which computes
the output observation set by simply reading it out from a
table look-up.   It is a model which the modeller has
available right at the beginning of the modelling exercise,
before he has found any structure or pattern in the system
behaviour.

For any system S, let the sets C_i, D_i be those defined
by def. (3.3.1).   One can think of the length of a concrete
(F,E)-model of S as the "perceived complexity", relative to
F, of the set $(C_1, \ldots, C_m)$, conditional on the set $((1, D_1), \ldots, (m, D_m))$.
The greatest lower bound of this "perceived complexity",
taken over all concrete (F,E)-models of S, is just the
conditional Kolmogorovcomplexity $K_F((C_1, \ldots, C_m) | ((1, D_1), \ldots, (m, D_m)))$.
(Although Kolmogorov complexity was developed for binary sequences
and binary programs, it can be readily generalised to sequences
and programs containing any finite number of symbols).   An
approximate upper bound for this Kolmogorov complexity is
the length of the trivial model of S.

The length of the trivial F model of S is the "perceived
complexity" of $(C_1, \ldots, C_m)$ before any structure has been

discovered in the system behaviour.   If a shorter model of

S is found, then its "perceived complexity" will be reduced.
Recalling the analogy between complexity and entropy, it
is appealing to measure the "perceived quantity of information"
in $((1,D_1),\ldots,(m,D_m))$ about $(C_1,\ldots,C_m)$ as the difference
between these two "perceived complexities".   Since
Kolmogorov complexity is not effectively computable, the only
upper bound on this "perceived quantity of information"
which is available, in general, is the length of the trivial
model.   Thus the length of the trivial model is a measure
of the amount of information potentially to be conveyed
by the modelling exercise.

Definition (3.4.2)

Let p be a concrete (F,E)-model of S, and let t be the
trivial F model of S.   Then the information gain I(p)
of p is the difference

$$I(p) = \ell(t) - \ell(p) \quad \ldots \ldots \ldots \ldots \ldots \ldots (3.5)$$

where $\ell(.)$ denotes the length of a program.

In section 1.1 a simple example was presented, which
suggested that the confidence which one has in a model
depends on the difference between the number of observations
which the model explains and the number of observations
required to construct the model.   The information gain is a
measure of this difference.   If the information gain is
zero, then all of the output observations have been used to
construct the model;   the trivial model is, of course, the
prime example of such a model.   If the information gain is

close to its upper bound $\ell(t)$, then the model is simple
enough to be constructed from only a small part of the output
observation set, and the remainder of this set is "explained"
by the model.   The information gain may be negative, of course.
This indicates that the model contains more arbitrary
"parameters" than is justified by the amount of information about
reality contained in the output observation set.

We are claiming, then, that the confidence in a model
increases as its information gain increases.   We assume, of
course, that all our knowledge about the system is contained
in the observation sets and in the definition of the programming
language (Chapter 4 deals with the latter aspect).   This claim
implies that the possible confidence we may have in some model
of a system is bounded by the size of the trivial model of
that system.   This accords well with the intuitive notion
that if we have only a few observations of a system, then
we can never have much confidence in any model of it which
may be postulated.

We embody our claim in the following axiom, on the basis
of which a choice may be made between competing models.

Axiom (3.4.3)

If S is a system, and E_1 and E_2 are sets such that
E_1- models of S and E_2- models of S are of interest, and p
and q are models, with p being an (F, E_1) -model of S and q
being an (F, E_2) -model of S, then the one of p and q which
has the higher information gain is to be chosen as the better
model of S.

This axiom implies that good models are small models.
Good models will therefore tend to use the same (short)
computational algorithm for as many computations as possible,
since the specification of every new algorithm increases
the size of the model.   Thus the above axiom provides a
specific link between the widely-held belief that simplicity
(as measured by smallness) is desirable in scientific hypotheses,
and the almost universal conviction that the more an observed
regularity has been repeated, the more likely it is to recur.

The following theorem is a crucial feature of our
characterisation of modelling.   As before, $\ell(p)$ denotes
the length of program p, measured by the number of terminal
characters which appear in it.

Theorem (3.4.4)

There is, in general, no effective procedure for finding
an (F,E)-model p of a system S, such that, for any other
(F,E)-model q of S, $\ell(p) \leqslant \ell(q)$.

Proof.   Suppose that such an effective procedure exists.
Consider the case $E=\{E_1\}=\{(\emptyset,Y)\}$ (where $S=(U,Y)$).   Then we
have an effective procedure for finding the shortest program
which computes Y, using only the set $\{1\}$.

Now suppose that the programming language F has only two
terminal characters.   Then there exists an effective
procedure for finding the shortest binary program which
computes Y, using $\{1\}$.   But Y can be associated uniquely with
a binary sequence: Y is a system, and can therefore be

associated with its index in some fixed enumeration of systems.
This index can be associated with a binary sequence by the
bijection introduced in section 2.2.1.   Since each of the
above steps is effective, there exists an effective procedure
for finding the shortest binary program which computes the
binary sequence associated with Y, and hence there exists
an effective procedure for finding its length, namely the
Kolmogorov complexity $K_F(Y|1)$.   Suppose F is optimal.
Then, by Church's Thesis, $K(Y|1)$ is partial recursive, which
contradicts theorem (2.2.5).   This proves the theorem.

Theorem (3.4.4) does not rely on F having only two
terminals or on E having the form indicated in the proof.
These assumptions are made in order to derive a contradiction
to theorem (2.2.5).   However, as mentioned earlier, this
theorem can be generalised to the case where the sequences
considered have an arbitrary finite number of symbols, and
to cover the uncomputability of conditional complexity.   On
the other hand, the theorem does rely on F being optimal.
A sufficiently restricted programming language may allow the
shortest model to be found, if it exists.   However, most
systems will not possess any models in such a language (the
simplest example is a programming language which always
computes the same thing, whatever program it may be given).

Theorem (3.4.4) implies that there is no algorithm for
finding the model of a system which has the highest information
gain.   So, according to our axiom, there is no algorithm for
finding the best model of a system.   Consequently the
modelling exercise cannot proceed according to some

"universal modelling algorithm", but must involve a process
of nonalgorithmic (creative?) postulation of hypotheses,
followed by the assessment of these hypotheses, according to
our axiom.

Note that the result of Meyer (28), which is mentioned
in section (2.2.4), implies that if a model is found, there
is still no algorithmic procedure for finding the shortest
implementation of that particular model.

3.5   Compatibility with Conventional Criteria.

Most models of data which has not been artificially
generated can be expected to contain at least one table look-
up, since it is most unlikely that every aspect of the data
can be explained.   The size of such a table look-up, as
measured by the mumber of characters required to program it,
can be regarded as a very general measure of prediction
error.   The bigger such a table is, the more features of
the data have not been explained by the model.

It will usually be possible to explain more of the
observed system behaviour - that is, to reduce the size of
any table look-ups in the model - only by using more elaborate
algorithms in the rest of the model - that is, by increasing
the size of the rest of the model.   Thus the use of smallness
of the program as a criterion of quality of a model leads to
a trade-off between the complexity of that part of the model
which would conventionally be regarded as the model (cf fig.2),
and the degree of approximation provided by that part of the

model to the observed data.   The use of this criterion
therefore provides a safeguard against "overfitting" the
model to the data.

If the output observation set is large enough, relative
to the size of that part of the model which is not a table
look-up, then the size of the table look-up(s) will dominate
the size of the model.   In this case, if two such models
are being compared, the use of the proposed criterion of
quality leads to the selection of the model with the smaller
table look-up(s).   This corresponds to the conventional
preference for small fitting errors, if the number of observations is large enough for the danger of overfitting to be
dismissed.

The definition of the programming language used determines the details of the trade-off implicit in the use of
the smallness criterion.   The manner in which table look-
up elements are coded, which constitutes part of the language
definition, is considered in Chapter 7.

A serious reservation must be made about the use of the
proposed criterion.   If sufficient _a priori_ knowledge
about the system is available then this may indicate that a
model which is not the smallest available should be preferred.
A typical example of this in conventional system identification
is the situation where _a priori_ knowledge indicates that a
parametric model with a particular structure is appropriate.
It may happen that a more elaborate model, when written as
a program, has a smaller overall size;   nevertheless, the
a priori knowledge will prevent that model being chosen as

better.   Another example is parameter estimation of a
linear dynamical process whose output is corrupted by noise.
In this case a straightforward minimisation of the equation
error usually leads to biased estimates (Eykhoff, (44)).
So if two models are being compared whose table look-ups
contain the equation errors, it is possible that the larger
one will be preferred on probabilistic grounds.   Once
again, a priori knowledge (about the noise) is required if
the smallness criterion is to be overridden.   Furthermore,
the smallness criterion could still be used to decide
between the larger of these two models and a third model
belonging to a different class.

As indicated in section 1.1, the proposed criterion is
intended for use in situations where little a priori information
is available, or in situations where it is too difficult to
use such a priori knowledge for model assessment.

The smallness - of -model criterion leads to the same
choice of model as do statistical considerations, for a very
important class of system behaviours and models of them.
If the system behaviour is a stationary random process with
rational spectral density function, then it is known how to
predict, at any time, its future behaviour, so as to minimise
the mean-square prediction error.   The method, due essentially
to Wiener and Kolmogorov, is to make the prediction for any
future time a suitable linear function of past observations
of the behaviour (45), (46).   If the observation intervals
are equally spaced, and a prediction is being made at each
instant of the system behaviour at the next observation instant,

then the prediction errors are equal to the random, uncorrelated disturbances which are imagined to be acting on the system.

Suppose it is desired to build a concrete (F,E) -model of the system which will give useful one-step-ahead predictions. Any E can be chosen which allows the model to use previous observations to compute predictions (cf. example (3.3.4)). The model will have to generate terms corresponding to prediction errors by means of a table look-up. If the programming language used codes table look-up terms in such a way that length of code is nondecreasing with the magnitude of the term (cf. Chapter 7), then, for a sufficiently long sequence of observations, the model with smallest (in magnitude) prediction errors will be the smallest (in length) model.

But it is known that, for the system under consideration, the smallest mean square prediction error is obtained by the use of the Wiener - Kolmogorov theory. Furthermore, Sherman (47) has shown that, if the process is Gaussian, then the same linear predictor is obtained if the expectation of any even nondecreasing function of the prediction error is minimised.

So, under these conditions, for a long enough sequence of observations, the "expected best model", judged according to axiom (3.4.3), is the Wiener - Kolmogorov model. The terms appearing in the table look-up of this model constitute a random, uncorrelated sequence. Theorem (2.2.12) therefore suggests that these terms could not be generated by any more efficient algorithm than a table look-up.

3.6 Prediction

If the best model that has been found up to some time is a concrete (F,E) -model p, and it is desired to find the system behaviour under some new (possibly not yet observed) conditions, which can be represented by a block of "virtual observations", D_{m+1}, that is, the observations which would be observed if the new conditions obtained, then the model p, and the computer F, can be used to find the "prediction" $F(p,m+1,D_{m+1})$. This provides a means of computing the values of a possible input/output function of the system on elements of its domain which have not been previously observed. According to our axiom, these values are the best "predictions" available to us. We have put the word "prediction" in quotes, because the value $F(p,m+1,D_{m+1})$ need not represent a future value (for example, if the model runs backwards through the observation interval).

It is possible, of course, that the value $F(p,m+1,D_{m+1})$ is not defined. In this case, it may not be possible to use p for predicting system behaviour under the conditions D_{m+1}. However, for some models, the value $F(p,m+1,D_{m+1})$ may be undefined simply because p includes the generation of certain parameters by means of a table look-up, and the table does not contain an element which is to be used for the (m+1)th computation. In this case, prediction is still possible if

such an element be supplied to the model.   The problem is,
what value should that element take?   Our solution is to
propose a second axiom, which may be thought of as an
extension of the previous one.

Axiom for Prediction

If elements are to be supplied to table look-ups of a
model, in order to allow that model to compute a prediction,
then the best prediction will be obtained if these elements
are chosen so as to minimise the resulting increase in
size of the model.

The use of this axiom must be qualified, of course, by
the value of the information gain.   We can supply an element
to a trivial model, thus enabling it to predict;   but we have
no confidence in that prediction, whatever the element may be.

A rough justification of the axiom runs as follows.
The basic assumption of scientific prediction is that any
regularities, which have been detected in a sequence of
observations, will continue to be present in the behaviour
of the system during the prediction interval.   Hence we
should choose the elements for the table look-up to be such
that previously observed regularities are present in the
computed prediction.   But, if the model is such that it
requires a large amout of code to appear in a table look-
up, in order to compute an output which is consistent with
such regularities, then we can certainly obtain a better model
by using the "fixed" part of the model (i.e. the part that

is common to all the computations) to compute the regularities.
This is true because for a sufficiently large set of
observations, the size of the model will be governed by the
average length of code appearing as elements of the table
look-up.   Thus the axiom is reasonable if it is assumed
that it is applied to the best available model.

The above argument can be illustrated by the following
example.   Suppose a system is defined by the observations:

$S=(U,Y)=(b,(552,553,546,551,549,544,547,554,557,551))$.

If the programming language and computing facility F is
taken to be Algol W, as implemented on the IBM 370/165
installation at Cambridge, and $E_i=(\emptyset,Y_i),i=1,...,10$, (so
that $E_3=(\emptyset,546)$, for example), then a trivial (F,E) -model
of S is:

```
      BEGIN INTEGER I,J;   INTEGER ARRAY Y(1::10);
            FOR J:=1 UNTIL 10 DO READ ( Y(J));
            READ (I);
            WRITE (Y(I));
END.
552,553,546,551,549,544,547,554,557,551,
```

When presented with an integer i ($1\leqslant i\leqslant 10$), this program
computes y_i by looking it up in the array Y.

We know that this model is useless for prediction,
because it is a trivial model.   Nevertheless, we can make
it compute a "prediction".   We must first supply it with a
new entry in its table look-up.   To do this, we replace the
integer 10 in line 2 by the integer 11, and add a new number

at the end of the program. When presented with the integer 11, this program will output the new number. What should this number be? According to our Axiom for Prediction, it should be one of the integers 0,...,9. The prediction of y_{11} will then be that integer.

Clearly this prediction is a very bad one. A prediction nearer 550 would "obviously" be better. In other words, we can see that a prediction obtained by not obeying the Axiom for Prediction will be better than one obtained by obeying it. But why can we see this? Because we have detected a regularity in the system behaviour-namely, that the behaviour tends to remain close to 550. But in that case we can use our knowledge of this to build a better model. One way of doing this is to observe that the mean of the behaviour remains close to 550, and therefore to build a model which is of the conventional "mean plus random error" type:

```
BEGIN INTEGER I,J;  INTEGER ARRAY  E(1::10);
      FOR J:=1 UNITL 10 DO READ (E(J));
      READ (I);
      WRITE (550+E(I));
END.
2,3,-4,1,-1,-6,-3,4,7,1,
```

This model computes the system behaviour Y by computing the observed regularity (550), and correcting it by a term obtained from a table look-up. Admittedly, this model is only slightly better than the trivial model (its information gain is 12 terminals, but it would rapidly become decisively

superior if more observations became available, which remained
close to 550.

In this case, if we apply our Axiom for Prediction, we
obtain as the predicted next output an integer lying between
500 and 559.   This time, the prediction is quite reasonable.

Clearly, several similar models can be built, and for
each of them there is a considerable range of"best" predictions.
It may be possible to reduce this range, for example by
estimating the probability distribution of the look-up
table terms.

3.7   An Example

3.7.1   Introduction

In this section an example will be presented, which
will portray a particular modelling exercise in terms of the
above characterisation.

The example used is the modelling of the gas furnace
data, which was considered by Box and Jenkins (45).   The
data, (which is given by Box and Jenkins as Series J),
consists of 296 pairs of input-output observations.   The
input observations are of gas flow rate into a furnace, and
the output observations are of the concentration of carbon
dioxide in the outlet gases.   The observations were made at
equal intervals of nine seconds.

Box and Jenkins obtain a model for these observations,
which consists of a deterministic transfer function relating
the input gas flow rate to the output concentration of carbon

dioxide, and a model of the noise process which disturbs the deterministic relationship. The model they obtain is:

$$\hat{y}_t = -\frac{0.53+0.37B+0.51B^2}{1-0.57B-0.01B^2} u_{t-3} \quad \ldots \ldots \ldots (3.6)$$

$$n_t = \frac{1}{1-0.53B+0.63B^2} w_t \ldots \ldots \ldots (3.7)$$

$$y_t^{\prime} = \hat{y}_t + n_t \ldots \ldots \ldots \ldots \ldots (3.8)$$

Here $u_t^{\prime}$ and $y_t^{\prime}$ represent the input and output variables, respectively, after removal of their mean values, at sampling instant t. $\hat{y}_t$ represents the estimate of $y_t^{\prime}$ generated by the transfer function of equn (3.6), and n_t is the error between $y_t^{\prime}$ and $\hat{y}_t$. Using conventional system identification terminology, n_t is an "output error" ("error in variables" in the terminology of econometrics (Johnston (48))). n_t represents a stochastic disturbance acting on the output of the process at time t, and w_t is a "discrete white-noise" process (i.e. a zero-mean, serially uncorrelated random sequence), which is considered to cause the disturbance n_t according to relationship (3.7). B is the backward shift operator, defined by $Bx_t = x_{t-1}$. The model can be given a conventional diagrammatic representation, as in fig. 3, where $\bar{u}, \bar{y}$ denote the mean values of the input and output variables, respectively.

3.7.2 The System

In terms of definition (1.3.1), the system which we are

considering is

$$S = (U, Y)$$

where $U = (u_1, \ldots, u_{296})$,

$\qquad Y = (y_1, \ldots, y_{296})$,

$\qquad \ell_i = m_i = 1$, for $i = 1, \ldots, 296$,

and the observations $\{u_i, y_i\}$ are as listed in Appendix C.

As in the example of section 3.6, we shall take the programming language F to be Algol W, as implemented on the IBM 370/165 installation at Cambridge.

3.7.3 Model I - The Trivial Model

We must define the sets A,B,C,D,E, which occur in definition 3.3.1. For the trivial model, we can take these to be:

$A = \{A_i : i = 1, \ldots, 296\}$, $A_i = \emptyset$

$B = \{B_i : i = 1, \ldots, 296\}$, $B_i = \emptyset$

$C = \{C_i : i = 1, \ldots, 296\}$, $C_i = y_i$

$D = \{D_i : i = 1, \ldots, 296\}$, $D_i = (A_i, B_i) = \emptyset$

$E = \{E_i : i = 1, \ldots, 296\}$, $E_i = (D_i, C_i) = (\emptyset, y_i)$

A concrete (F,E) -model, which is a trivial model of the system S, is:

```
BEGIN  INTEGER I,J;  REAL ARRAY Y(1::296);
       FOR J:=1 UNTIL 296 DO READON (Y(J));
       READ (I);
       WRITE (Y(I));
END.
```

53.8   53.6   53.5   . . .   57.0

The last line of the trivial model is the table look-up, which contains the output observations. The trivial model can be represented diagrammatically, as in fig. 4(a).

3.7.4 Model II - The Mean

Probably the first nontrivial model to be hypothesised for many systems is that the system behaviour has a constant mean value. This model is of the type which reproduces regularities only in the output observations, and does not exploit any information in the input observations. Consequently, the sets A,B,C,D,E may be taken to be the same as for the trivial model. The mean value of the output observations is 53.5. The following is a (P,E)-model of S which makes use of this fact:

```
BEGIN   INTEGER I,J;   REAL ARRAY Y(1::296);
        FOR J:=1 UNTIL 296 DO READON (Y(J));
        READ (I);
        WRITE  (53.5 + Y(I));
END.
```

.3   .1   0   0   -.1   ...   3.8   3.5

The table look-up of this model is listed in the column headed y_i' in Appendix C. Fig. 4(b) shows a diagrammatic representation of this model. The dashed line represents the boundary of the model.

3.7.5 Model III - Deterministic Transfer Function

We now assume that the transfer function of equation

(3.6) has been hypothesised as the relationship between
the input and output of the system.   However, we make the
restriction that the model may not assume knowledge of past
output observations (other than the initial conditions), but
may assume knowledge of all the past and present input
information.   We have to choose new sets A, ..., E:

$$A = \{A_i : i=1,...,296\}, \quad \begin{cases} A_i = (u_1, u_2, ..., u_i) \text{ for } i \geq 6 \\ A_i = \emptyset \text{ for } i < 6 \end{cases}$$

$$B = \{B_i : i=1,...,296\}, \quad B_i = \emptyset$$

$$C = \{C_i : i=1,...,296\}, \quad C_i = Y_i$$

$$D = \{D_i : i=1,...,296\}, \quad D_i = (A_i, B_i) = (u_1, u_2, ..., u_i)$$

$$E = \{E_i : i=1,...,296\}, \quad E_i = ((u_1, u_2, ..., u_i), Y_i).$$

An (F,E) -model, which uses the hypothesis that the system
behaviour is governed by equation (3.6), is:

```
BEGIN   INTEGER I,J;   REAL ARRAY N,U,Y(1::296);
        FOR J:=1 UNTIL 296 DO READON (N(J));
        READ (I);
        IF I<6 THEN WRITE (N(I)) ELSE
BEGIN
        FOR J:=1 UNTIL 5 DO
        BEGIN
            Y(J):= N(J)-53.5;
            READON (U(J));
        END;
        FOR J:=6 UNTIL I DO
```

```
        BEGIN

            READON (U(J));

            Y(J):=-.53*U(J-3)-.37*U(J-4)-

                .51*U(J-5)+.57*Y(J-1)+

                    .01*Y(J-2);

        END;

        WRITE (Y(I)+53.4+N(I));

    END;

END.
```

53.8   53.6   53.5   53.4   -.2   -.4   ...   4.1   4.0

The table look-up of this model is listed in Appendix C in the column headed n_i. Note that the first five terms of the table are simply the output observations $y_1, \ldots, y_5$. The reason for this is that equation (3.6) relates the value of y_i to the values of $u_{i-3}, u_{i-4}, u_{i-5}$. Hence this equation cannot be used for generating the first five terms of the observed behaviour. So the first five terms are generated by table look-up. Fig. 4(c) shows a representation of this model. The sequence $\{n_i\}$ is the same as the sequence $\{n_t\}$ defined by equation (3.8).

Note that fig. 4(c) shows the subtraction of the mean of the input observations before these observations are submitted to the transfer function algorithm. In fact, the model achieves this more economically by having 53.4 instead of 53.5 in its output statement.

3.7.6 Model IV - Deterministic Transfer Function, Using Output Observations.

Now suppose that the same transfer function is hypothesised as governing the behaviour of the system, but that the model is now allowed to use all past output observations, as well as past and present input observations. Because of the nature of the transfer function, suitable sets A,...,E are:

$$A = \{A_i : i=1,\ldots,296\}, \quad A_i=\emptyset \text{ for } i<6, \ A_i=(u_{i-5},u_{i-4},u_{i-3})$$
$$\text{for } i\geqslant 6,$$

$$B = \{B_i : i=1,\ldots,296\}, \quad B_i=\emptyset \text{ for } i<6, \ B_i=(y_{i-1},y_{i-2})$$
$$\text{for } i\geqslant 6,$$

$$C = \{C_i : i=1,\ldots,296\}, \quad C_i=y_i$$

$$D = \{D_i : i=1,\ldots,296\}, \quad D_i=(A_i,B_i) \begin{cases} =\emptyset \text{ for } i<6 \\ =((u_{i-5},u_{i-4},u_{i-3}),(y_{i-1},y_{i-2})) \end{cases}$$
$$\text{for } i\geqslant 6,$$

$$E = \{E_i : i=1,\ldots,296\}, \quad E_i=(D_i,C_i) \begin{cases} =(\emptyset,y_i) \text{ for } i<6 \\ =(((u_{i-5},u_{i-4},u_{i-3}),(y_{i-1},y_{i-2}), \\ \qquad\qquad\qquad\qquad\qquad y_i) \end{cases}$$
$$\text{for } i\geqslant 6.$$

A suitable (F,E) -model is:

```
BEGIN  INTEGER I,J;  REAL ARRAY E(1::296);  REAL U,V,W,Y,Z;
       FOR J:=1 UNTIL 296 DO READON (E(J));
       READ (I);
       IF I<6 THEN WRITE (E(I)) ELSE
       BEGIN
            READON (U,V,W,Y,Z);
```

WRITE (-.53*U-.37*V-.51*W+.57*Y + .01*Z

+ 22.4 + E(I));

END;

END.

53.8   53.6   53.5   53.5   53.4   -.2   -.3   ...   1.7   1.6

The table look-up for this model is listed as column e_i

in Appendix C.   The data items required by this model are:

$U=u_{i-3}$, $V=u_{i-4}$, $W=u_{i-5}$, $Y=y_{i-1}$, $Z=y_{i-2}$.   The term "22.4", which

appears in the output statement, corrects for the non-zero

means of both input and output observations.   A diagrammatic

representation of this model is shown in fig. 4(d).

Note that the elements e_i in the table look-up are not

the same as the terms n_t which appear in equation (3.8).

The reason for this is, of course, that the output of the

transfer function part of the model is no longer $\hat{y}_i$, but is

a new quantity y_i^*. $\hat{y}_i$ is given by

$$\hat{y}_i = -0.53u'_{i-3} - 0.37u'_{i-4} - 0.51u'_{i-5} + 0.57\hat{y}_{i-1} + 0.01\hat{y}_{i-2}. \quad . \quad (3.9)$$

whereas y_i^* is given by

$$y_i^* = -0.53u'_{i-3} - 0.37u'_{i-4} - 0.51u'_{i-5} + 0.57y'_{i-1} + 0.01y'_{i-2}. \quad . \quad (3.10)$$

Since $y'_i = \hat{y}_i + n_i$, we have

$$y_i^* = \hat{y}_i - 0.57n_{i-1} - 0.01n_{i-2} \quad . \quad . \quad . \quad . \quad . \quad . \quad . \quad . \quad . \quad . \quad (3.11)$$

In general, $y_i^* = \hat{y}_i + (1-A(B))n_i \quad . \quad . \quad . \quad . \quad . \quad . \quad . \quad . \quad . \quad . \quad (3.12)$

if the scalar term of A(B) is 1, where A(B) is defined as in fig. 3.

Eqns (3.8) and (3.12), together with

$$e_i = y'_i - y_i^* \quad . \quad . \quad . \quad . \quad . \quad . \quad . \quad . \quad . \quad . \quad . \quad . \quad (3.13)$$

lead to $\quad e_i = A(B)n_i$ . . . .(3.14)

which is a well-known result (44). e_i is in fact the "equation error" for the model $y' = \frac{B}{A}u'$.

In view of these differences, model IV is not of the form which was assumed by Box and Jenkins, and for which the model coefficients were estimated. This does not prevent us from measuring its information gain, and thus obtaining some assessment of its value as a one-step-ahead predictor which uses the most recent output observations. This underlines an important feature of our characterisation of modelling. Theorem (3.4.4) implies that, for the very general class of models which we consider (those of definition (3.3.1)), there exists a fundamental dichotomy between finding a model and assessing it. Thus the process by which a particular model was arrived at is quite irrelevant to its assessment by the use of information gain.

3.77   Model V - Stochastic Process Model

We now consider a refined version of model IV, which uses eqn (3.7) in an attempt to predict the terms e_i. From a statistical point of view this is a nonsense, because the coefficients appearing in eqn (3.7) were estimated for the process n_i, and eqn (3.14) shows that the processes e_i and n_i have quite different spectral characteristics. Once again, however, we are free to assess the value of the model for one-step-ahead prediction, regardless of how it was obtained.

In this case the model must be allowed access to past output observations, since it would otherwise be unable to exploit equation (3.7). However, the sets $A, \ldots, E$ must be slightly different from those defined for model IV:

$A = \{A_i : i = 1, \ldots, 296\},$ $\qquad A_i = \emptyset$ for $i < 8,$ $\quad A_i = (u_{i-7}, \ldots, u_{i-1})$

$$\text{for } i \geqslant 8,$$

$B = \{B_i : i = 1, \ldots, 296\},$ $\qquad B_i = \emptyset$ for $i < 8,$ $\quad B_i = (y_{i-7}, \ldots, y_{i-1})$

$$\text{for } i \geqslant 8,$$

$C = \{C_i : i = 1, \ldots, 296\},$ $\qquad C_i = y_i,$

$D = \{D_i : i = 1, \ldots, 296\},$ $\qquad D_i = (A_i, B_i) = \begin{cases} \emptyset & \text{for } i < 8 \\ ((u_{i-7}, \ldots, u_{i-1}), (y_{i-7}, \ldots, y_{i-1})) \\ \qquad \text{for } i \geqslant 8, \end{cases}$

$E = \{E_i : i = 1, \ldots, 296\},$ $\qquad E_i = (D_i, C_i) = \begin{cases} (\emptyset, y_i) & \text{for } i < 8 \\ (((u_{i-7}, \ldots, u_{i-1}), (y_{i-7}, \ldots, \\ \quad y_{i-1})), y_i) & \text{for } i \geqslant 8. \end{cases}$

A model which exploits equations (3.6) and (3.7) could be built for smaller sets A_i and B_i than those defined above. However, the model would then have to be slightly larger. Since real interest lies not in building an (F,E)-model for a particular set E, but rather for any nonanticipative E (cf. definition (3.3.5)), it is sensible to define the sets A, $\ldots$,E in a way that allows the smallest (F,E)-model to be built, providing that E remains nonanticipative.

In this case the (F,E)-model is:

```
BEGIN    INTEGER I,J;   REAL ARRAY A,U,Y,Z (1::296);
         FOR J:= 1 UNTIL 296 DO READON (A(J));
         READ (I);
         IF I<8 THEN WRITE (A(I)) ELSE
```

```
        BEGIN
                    FOR J:=1 UNTIL 7 DO READON (U(I-J),Y(I-J));
                    FOR J:=(I-2) UNTIL I DO
                    Z(J):=-.53*U(J-3)-.37*U(J-4)-.51*U(J-5)
                         +.57*Y(J-1)+.01*Y(J-2);
                    WRITE (Z(I)+1.53*(Y(I-1)-Z(I-1))-.63*(Y(I-2)-Z(I-2))
                    + A(I)+2.2);
        END;
END.
```

53.8   53.6   53.5   53.5   53.4   53.1   52.7   .1   0´....   .2

The table look-up for this model is listed as column a_1
in Appendix C.   All the required adjustments to the means of
the observations are accomplished by the single term "2.2"
in the output statement.   A representation of the model is
shown in fig. 4(e).

The above form of the model shows clearly how the output
observations are computed, but is slightly larger than
necessary.   A shorter version which performs the same
computation is:

```
        BEGIN
                    INTEGER I,J;   REAL ARRAY A,U,Y(1::296);
                    FOR J:=1 UNTIL 296 DO READON (A(I));
                    READ (I);
                    IF I<8 THEN WRITE (A(I)) ELSE
        BEGIN
                    FOR J:=1 UNTIL 7 DO READON (U(I-J),Y(I-J));
                    WRITE (2.1*Y(I-1)-1.5*Y(I-2)+.34*Y(I-3)
                         +.01*Y(I-4)-.53*U(I-3)+.44*U(I-4)
```

```
                    -.28*U(I-5)+.55*U(I-6)-.32*U(I-7)
                    +A(I)+2.2);
        END;
END.
53.8              53.6              ...              .2
```

The table look-up for this version is of course the same as
for the previous one.

3.7.8 Model VI - Box & Jenkins Model

Finally, we consider the model which uses equations
(3.6), (3.7) and (3.8) in the manner intended by Box and
Jenkins (45).   The sets A,...,Ecan remain the same as for
model V.   The model, which can be compared with the forecast
function given on p.407 of (45), is:

```
        BEGIN
                INTEGER I,J;   REAL ARRAY  W,U,Y(1::296);
                FOR J:=1 UNTIL 296 DO READON (W(J));
                READ (I);
                IF I<8 THEN WRITE (W(I)) ELSE
                BEGIN
                        FOR J:=1 UNTIL 7 DO READON (U(I-J),Y(I-J));
                        WRITE (2.1*Y(I-1)-1.5*Y(I-2)+.34*Y(I-3)
                               +.01*Y(I-4)-.53*U(I-3)+.44*U(I-4)
                               -.28*U(I-5)+.55*U(I-6)-.32*U(I-7)
                               +W(I)-.57*W(I-1)-.01*W(I-2)+2.2);
        END;
END.
53.8   53.6   53.5   53.5   53.4   53.1   52.7   .1   .1   ...   .4
```

The table look-up for this model is shown in Appendix C as the column headed w_i. The elements of this column are estimates of the elements of the "white noise" sequence $\{w_t\}$ of equation (3.7).

Fig. 4(f) shows the structure of this model, although the above program is a more efficient implementation than that shown in the figure (cf model V). The equivalence of this model to equations (3.6)-(3.8) is shown by the following manipulations, where the operators A,B,C,D are as defined in fig. 3:

$$y_i' = y_i{}^* + e_i$$

$$y_i{}^* = (1-A)y_i' + Bu_i'$$

$$e_i = (1-C)e_i + ADw_i$$

$$= (1-C)(y_i' - y_i{}^*) + ADw_i$$

$$= (1-C)(Ay_i' - Bu_i') + ADw_i .$$

$$\therefore y_i' = (1-AC)y_i' + BCu_i' + ADw_i$$

$$\therefore ACy_i' = BCu_i' + ADw_i$$

$$\therefore y_i' = \frac{B}{A}u_i' + \frac{D}{C}w_i .$$

Note that the operators (1-A) and (1-C) act only on past values of y_i' and e_i, respectively.

3.7.9 Assessment of the Models

The size of each of the above six models was measured as the number of terminal characters of Algol W that appears in it. Reserved words, such as BEGIN, were considered to be single terminals, as were standard procedure names, such

as WRITE.   This practice is justified in chapter 6.
Unnecessary spaces were not counted.   The elements of each
table look-up were taken to be as shown in Appendix C,
except that positive entries were considered to be preceded
by "+".   The reason for this is discussed in chapter 7.

The following table gives the results of the model
assessment.   In addition to the size and information gain
of each model, the "information explained" by it is shown.
This quantity is the ratio of the information gain to the
size of the trivial model and resembles an efficiency measure,
if the information gain is not negative.

MODEL	SIZE	INFORMATION GAIN	INFORMATION EXPLAINED
I	1532	O	O
II	1159	373	24.4%
III	1076	456	29.8%
IV	964	568	37.0%
V	1005	527	34.4%
VI	1013	519	33.8%

TABLE I - Results of Model Assessment.

It is interesting to consider what the situation would be if
the same six models were postulated for systems which
consisted of the initial segments of the data, $S_j=((u_1,\ldots,u_j),$
$(y_1,\ldots,y_j))$, $0 \leqslant j \leqslant 296$.   The model sizes and information gains
of the six models for these systems are shown in figs.5
and 6, respectively.

From fig. 6 it is seen that after the first few
observations become available, the best of the six models

is model II - namely, the postulation of a constant mean.
This model remains the best until about 90 observations have
been obtained, whereupon model IV becomes better. Thereafter
model IV remains the best of the five models. Model V - a
more elaborate model than model IV - has a lower information
gain than model IV, even after all 296 observations have
been obtained. Examination of Appendix C reveals that model V
does actually have smaller prediction errors than model IV.
Nevertheless, the comparison of information gains indicates
that the increase in complexity from model IV to model V
is not justified; the model is being "overfitted" to the data.

Model III has a lower information gain than model V.
This is not very surprising, since model V is predicting one-
step-ahead, using the latest output information at each step,
whereas model III is predicting over the whole observation
interval, using only the input information. However, model
III has a higher information gain than model II. This
indicates that for long-term prediction it is better to use
model III, rather than just predict the mean value of the
output observations.

Model VI also has a slightly lower information gain than
model V. The prediction errors w_i are slightly smaller, on
the whole, than those of model V, but the difference is too
insignificant to justify the increased complexity of model
VI. It is interesting to note that, although model VI is
of the form intended by Box and Jenkins, while model V is
not, there is little to choose between them on the basis of
information gain. Furthermore, neither model is preferable
to model IV.

3.8   Summary

The characterisation of modelling which has been developed
in this chapter can be summarised as follows:

(1)   A system is defined by a set of observations.

(2)   A model of a system is an algorithm for computing
the output observation set by using specified subsets of
the system observations.

(3)   Those aspects of a system's behaviour which are
not understood are computed by the model with the aid of a
table look-up.

(4)   The situation at the beginning of the modelling
exercise is captured by the concept of the trivial model.

(5)   The modelling exercise progresses in "steps",
each step resulting from the postulation of hypotheses about
the system.   In general, the transition from one step to
the next cannot proceed algorithmically.

(6)   At each step, the progress of the modelling
exercise is measured by the information gain of the current
model.

4. INCORPORATION OF A PRIORI KNOWLEDGE

4.1 Choice of Programming Language.

The modeller has certain a priori beliefs about the
system he is modelling. His choice of programming language
should reflect these. It will be recalled from sec. 3.4
that assessing models on the basis of information gain is
tantamount to comparing the number of "arbitrary elements"
which make up a hypothesis about a behaviour with the "number
of observations" of that behaviour. These arbitrary elements
are always counted relative to some structure which is taken
for granted. This structure is provided by the definition
of the programming language used. In other words, the
programming language embodies those arbitrary elements which
will be common to all the models being assessed. It
obviously makes sense to choose the language so that these
common elements coincide with those assumptions that the
modeller is willing to take for granted.

For example, suppose that the modeller believes that
the gas-furnace data of sec. 3.7 is certainly produced by
a model of the form

$$y_i = b_0 u_i + b_1 u_{i-1} + \ldots + b_m u_{i-m} - a_1 y_{i-1} - \ldots - a_n y_{i-n} + e_i \quad \ldots \ldots \quad (4.1)$$

and that he is not prepared to seriously consider a model
with any other structure. Then he can (informally) define
the following programming language. Every program of the
language is a list of integers and rationals which is given
the interpretation:

$$n, m, a_1, \ldots, a_n, b_0, \ldots, b_m, e_1, \ldots, e_N.$$

The data for such a program is a similar list, with the
interpretation:

$$i, y_{i-1}, \ldots, y_{i-n}, u_i, \ldots, u_{i-m}.$$

Given such a program and such a set of data, the computation
of y_i in accordance with eqn (4.1) is invoked.   If input
observations $(u_1, \ldots, u_N)$ and output observations $(y_1, \ldots, y_N)$
of a system are obtained, then a certain (infinite) set of
programs in this language will constitute models of the system
$((u_1, \ldots, u_N), (y_1, \ldots, y_N))$.   The terms $e_1, \ldots, e_N$ which appear
in the program form a table look-up.   A trivial model is
obtained if $m=n=b_0=0$, and $e_1=y_1, \ldots, e_N=y_N$.

The programming language just described will be called
Linear Model Language, or LML.   Fig. 7 shows the structure
of the computations it performs.   LML is used in Appendix
A to illustrate the formal definition of programming languages.

Assuming that the same representation of numbers is used
in LML as in Algol, it is clear that a model written in LML
will be smaller than the Algol implementation of the same
algorithm.   So model assessment using LML will indicate
fewer "arbitrary elements" in each model than would assessment
using Algol.   In a sense, some of the arbitrary elements
have been shifted from the definition of each program to the
definition of the language.   This is seen clearly if LML is
considered to be an Algol procedure, rather than a separate
language.

It will be seen later in this chapter that the choice
of programming language can affect the results of model
assessment.   So, if the choice of language is considered

to be the specification of the modeller's _a priori_ knowledge,
then model assessment on the basis of information gain is
seen to depend on _a priori_ knowledge.   This is a feature of
any method of model assessment, and is not specific to the
method being advocated here.

The modeller will often be uncertain about the correctness
of his _a priori_ beliefs.   Fortunately, he does not always
have to choose between conflicting assumptions.   He can
choose between programming languages which imply a greater
or lesser state of knowledge.   For example, the choice
of LML for model assessment implies much more specific
knowledge about the system than does the choice of Algol.
An intermediate state of knowledge may perhaps be represented
by use of a simulation language.

Note that the choice of language is not just a choice
between a special-purpose and a universal language.   The
modeller's _a priori_ beliefs may coincide fairly well with
the structure embodied in some subset of Algol, but may be
quite different from that embodied in some language designed
for manipulating strings.   Nevertheless both languages may
be universal, in the sense that each is capable of computing
every partial recursive function.

An obvious restriction which must be placed on a language
which is to be used for model assessment arises as follows.
Suppose that the language being used includes a standard
procedure which can be called by its single-terminal identifier,
say A.   Suppose further that this procedure computes the
output observation set of the system S by means of a table

look-up.   Then a program consisting of little more than the single terminal A would be a model of S.   This model would be a shorter, and hence a better, model of S than any other model written in that language, yet it could be constructed without any understanding of the system S.   Clearly, this situation must be outlawed.

Insisting on the inadmissibility of such a procedure can be regarded as another aspect of specifying the modeller's *a priori* knowledge about the system.   The use of the procedure would imply that the system behaviour is not problematic − that the modeller is willing to accept the system behaviour as an "exogenous variable" of the world which he is examining, and one which he does not wish to investigate.   Such an acceptance would, of course, render the modelling exercise redundant.

4.2   Asymptotic Models

In this section it is shown that the choice of programming language does not affect model assessment if the observation sets are large enough.   The characterisation of modelling developed in chapter 3 considered a system to be defined by a single pair of observation sets.   We now wish to extend this, so as to describe the modelling of an increasing set of observations.

Definition (4.2.1)

(a)   Let $U_1 = (u_1, \ldots, u_i)$ and $U_2 = (u_{i+1}, \ldots, u_j)$ be observation sets.   Then the observation set

$U_1 U_2 = (u_1, \ldots, u_i, u_{i+1}, \ldots, u_j)$ is an <u>extension</u> of U_1.

(b)   Let $S_1 = (U_1, Y_1)$ and $S_2 = (U_2, Y_2)$ be systems.   Then

the system $S_1 S_2 = ((U_1 U_2), (Y_1 Y_2))$ is an <u>extension</u> of S_1.

<u>Definition (4.2.2)</u>

An infinite sequence of systems
$$\mathcal{S} = (S^1, S^2, \ldots)$$
where S^j is an extension of S^{j-1} for every $j > 1$, (i.e.
$S^j = S_1 S_2 \ldots S_j$) is an <u>asymptotic system</u>.

We wish to consider models of the S^j's which differ
only in their table look-ups.   To capture the idea of a
table look-up without restricting it unduly, we shall
consider models to be pairs (m, T).   Each element of the pair
is a part of a program, and the pair (m, T) is regarded as
the complete program.   This can be formalised quite easily,
if required: take a pairing function τ (cf proof of theorem
(3.2.1)), and change definition (3.3.6), so that a concrete
(F, E)-model becomes an ordered pair of integers (m, T), such
that $F(\tau(m, T), i, D_i) = C_i$.   These integers can be associated
with programs, as before. m will be considered to be the
part which is common to models of all the S^j's, while T^j
will be regarded as a table look-up, which may be different
for each S^j.   When a translation of the program (m, T) from
one language to another is considered, it will be assumed that
T (or at least its length) remains unchanged.   On the other
hand, the translation of m will be assumed to be different

from m.   In this way a distinction is drawn between T and m, which corresponds to some aspects of the distinction between table-lookup and other types of program.

In the following definition a particular programming language is assumed.   m and T^j are fragments of programs in this language.   The definition is based on definition (3.3.1), and the notations of definition (4.2.1) are generalised in an obvious manner.

Definition (4.2.3)

Let $A^j = \{A^j_i\}$ be a set of ordered subsets of $(U_1 U_2 \ldots U_j)$, let $B^j = \{B^j_i\}$ be a set of ordered subsets of $(Y_1 Y_2 \ldots Y_j)$, and let $C^j = \{C^j_i\}$ be a complete set of m_j disjoint ordered subsets of $(Y_1 Y_2 \ldots Y_j)$.   Let D^j be a set of ordered pairs $D^j_i = (A^j_k, B^j_\ell)$ $(i=1,2,\ldots,m_j)$, and let E^j be a set of ordered pairs $E^j_i = (D^j_i, C^j_i)$ $(i=1,2,\ldots,m_j)$.   Finally, let ξ be the sequence $\xi = (E^1, E^2, \ldots)$, and $\mathcal{T}$ be the sequence $\mathcal{T} = (T^1, T^2, \ldots)$.

Then the pair $(m, \mathcal{T})$ is an __asymptotic ξ-model__ of the asymptotic system $\mathcal{S} = (S^1, S^2, \ldots)$ if and only if (m, T^j) is an E^j-model of S^j, for every $j=1,2,\ldots$

The following definitions distinguish between two possible asymptotic behaviours of rival models. $I(m, T^j)$ denotes the information gain of the model (m, T^j), and $E(m, T^j)$ denotes the information explained by model (m, T^j), namely the ratio of $I(m, T^j)$ to the size of the trivial model of S^j. $(m_1, \mathcal{T}_1)$ and $(m_2, \mathcal{T}_2)$ denote asymptotic models of some asymptotic system $\mathcal{S}$, with $\mathcal{J}_1 = (T_1^1, T_1^2, \ldots)$ and $\mathcal{J}_2 = (T_2^1, T_2^2, \ldots)$.

We use $\lim\inf_{j \to \infty} x_j$ to denote $\lim\inf_{j \to \infty} \underset{k > j}{} x_k$, and

similarly for lim sup.

Definition (4.2.4)

$(m_1, \mathcal{J}_1)$ is <u>asymptotically weakly better</u> than $(m_2, \mathcal{J}_2)$
(denoted by $(m_1, \mathcal{J}_1) >_w (m_2, \mathcal{J}_2)$) if and only if

$$\lim\inf_{j \to \infty} \{I(m_1, T_1^j) - I(m_2, T_2^j)\} = +\infty. \quad\ldots\ldots\ldots\ldots \quad (4.2)$$

Definition (4.2.5)

$(m_1, \mathcal{J}_1)$ is <u>asymptotically strongly better</u> than $(m_2, \mathcal{J}_2)$
(denoted by $(m_1, \mathcal{J}_1) >_s (m_2, \mathcal{J}_2)$) if and only if

$$\lim\inf_{j \to \infty} \{E(m_1, T_1^j) - E(m_2, T_2^j)\} > 0 \quad\ldots\ldots\ldots\ldots \quad (4.3)$$

The ideas behind these definitions are the following.
Let t_j denote the trivial model of S^j, and $|t_j|$ denote its
size. We henceforth make the natural assumption that

$$\lim_{j \to \infty} |t_j| = +\infty \quad\ldots\ldots\ldots\ldots\ldots\ldots\ldots\ldots \quad (4.4)$$

If $(m_1, \mathcal{J}_1)$ is asymptotically weakly better than $(m_2, \mathcal{J}_2)$ then

the "amount of information" extracted from S^j by (m_1, T_1^j) is

eventually greater than that extracted by (m_2, T_2^j), and the

difference between them is eventually increasing. But

their "rates of information extraction", as measured by the

information explained, may be converging towards each other.

For example, if $|t_j|=kj$, $I(m_1,T_1^j)=pj^{\frac{1}{2}}$, $I(m_2,T_2^j)=qj^{\frac{1}{2}}$, with $p>q$,

then $I(m_1,T_1^j)-I(m_2,T_2^j)=(p-q)j^{\frac{1}{2}}\to\infty$, while

$E(m_1,T_1^j)-E(m_2,T_2^j)=\frac{p-q}{k}j^{-\frac{1}{2}}\to 0$.

If $(m_1,\mathcal{T}_1)$ is asymptotically strongly better than

$(m_2,\mathcal{T}_2)$ then the "rate of information extraction" by $(m_1,\mathcal{T}_1)$

is eventually greater than that by $(m_2,\mathcal{T}_2)$. The "weak/

strong" terminology is justified by the following theorem.

<u>Theorem (4.2.6)</u>

$$(m_1,\mathcal{T}_1)>_s(m_2,\mathcal{T}_2)\Longrightarrow(m_1,\mathcal{T}_1)>_w(m_2,\mathcal{T}_2).$$

<u>Proof</u>

Suppose $\lim_{j\to\infty}\inf\{I(m_1,T_1^j)-I(m_2,T_2^j)\}<+\infty$.

Then $\exists N$, such that for any integer k, $\exists i>k$, such that

$$I(m_1,T_1^i)-I(m_2,T_2^i)\leqslant N.$$

Since $E(m,T^i)=\frac{I(m,T^i)}{|t_i|}$ and $|t_i|\to\infty$, this implies that for any

integer k, and for any $\varepsilon>0,\exists i>k$, such that

$$E(m_1,T_1^i)-E(m_2,T_2^i)<\varepsilon .$$

But this contradicts $\lim_{j\to\infty}\inf\{E(m_1,T_1^j)-E(m_2,T_2^j)\}>0$.

Hence $\lim_{j\to\infty}\inf\{E(m_1,T_1^j)-E(m_2,T_2^j)\}>0\Longrightarrow\lim_{j\to\infty}\inf\{I(m_1,T_1^j)-$

$I(m_2,T_2^j)\}=+\infty$.

We now consider the effect of writing models in different

languages on their asymptotic performance. For a precise

discussion of what it means for a program to be written in

some particular language, see chapter 5. Let $(m_1, \mathcal{J}_1)$ and
$(m_2, \mathcal{J}_2)$ be asymptotic models of $\mathcal{S}$ written in a programming
language μ. Let π be a programming language, such that
programs $(p_1, T_1^j), (p_2, T_2^j)$, $j=1,2,\ldots$, can be written in π,
and such that these programs compute the same partial
recursive functions as the programs $(m_1, T_1^j), (m_2, T_2^j)$, $j=1,2,\ldots$,
respectively. Using the notation of definition (3.3.6)
we can write

$$\pi(\tau(p_1, T_1^j), \cdot, \cdot) = \mu(\tau(m_1, T_1^j), \cdot, \cdot) \quad \ldots \ldots \quad (4.5)$$

where τ is an appropriate pairing function, and similarly
for p_2, m_2. Consequently $(p_1, \mathcal{J}_1)$ and $(p_2, \mathcal{J}_2)$ are asymptotic
models of $\mathcal{S}$ written in π.

Let $|p|$ denote the size of a program p; let t_j^μ be the
trivial model of s^j written in μ, and let t_j^π be the trivial
model of s^j written in π. We assume that

$$|t_j^\pi| = |t_j^\mu| + k_1 \quad \ldots \ldots \ldots \ldots \ldots \quad (4.6)$$

Theorem (4.2.7)

With the notations and assumptions as stated above,

(a) $(m_1, \mathcal{J}_1) >_w (m_2, \mathcal{J}_2) \Leftrightarrow (p_1, \mathcal{J}_1) >_w (p_2, \mathcal{J}_2)$

(b) $(m_1, \mathcal{J}_1) >_s (m_2, \mathcal{J}_2) \Leftrightarrow (p_1, \mathcal{J}_1) >_s (p_2, \mathcal{J}_2)$

Proof

(a)   There exist integers k_2, k_3, such that

$$|p_1| = |m_1| + k_2 \text{ and } |p_2| = |m_2| + k_3.$$

Hence $\quad I(m_1,T_1^j) - I(m_2,T_2^j) = |m_2|+|T_2^j|-|m_1|-|T_1^j|$

$$= |p_2|+|T_2^j|-|p_1|-|T_1^j|+k_2-k_3$$

$$= I(p_1,T_1^j)-I(p_2,T_2^j)+k_2-k_3 \quad .$$

The result follows from definition (4.2.4).

(b) $\quad \lim_{j\to\infty} \inf \{E(p_1,T_1^j)-E(p_2,T_2^j)\}=$

$$= \lim_{j\to\infty} \inf \frac{1}{|t_j^\pi|}\{I(p_1,T_1^j)-I(p_2,T_2^j)\}$$

$$= \lim_{j\to\infty} \inf \frac{\{I(m_1,T_1^j)-I(m_2,T_2^j)\}-k_2+k_3}{|t_j^\mu|} \quad . \quad \frac{|t_j^\mu|}{|t_j^\mu|+k_1}$$

$$= \lim_{j\to\infty} \inf \{E(m_1,T_1^j)-E(m_2,T_2^j)\} \quad , \text{ by eqn (4.4)}.$$

The result follows from definition (4.2.5).

Theorem (4.2.7) shows that the asymptotic relative merits of two asymptotic models are not changed by a change of programming language. This result, coupled with the view that choice of programming language specifies a priori knowledge, has the following interpretation: According to our characterisation of modelling, the relative merits of two rival models are independent of the modeller's a priori beliefs, if the observation sets are large enough. This is a weak condition which should be satisfied by any reasonable procedure for model assessment.

4.3   Practical Effect of Change of Language.

The asymptotic results of theorem (4.2.7) say nothing about the situation for small observation sets. In this section it is demonstrated that a change of programming language can affect the results of model assessment in practice.

Models I, II, IV and V of the gas-furnace observations (cf. section 3.7) will be assessed using Extended Linear Model Language (ELML). This is a language similar to LML, which was described in section 4.1, but which performs computations of the form

$$y_i = b_o u_i + \ldots + b_m u_{i-m} - a_1 y_{i-1} - \ldots - a_n y_{i-n}$$

$$e_i = d_o w_i + \ldots + d_p w_{i-p} - c_1 e_{i-1} - \ldots - c_q e_{i-q}$$

$$y_i = \hat{y}_i + e_i + \bar{y} \tag{4.7}$$

Fig. 8 shows the structure of these computations. Each ELML program is a list of integers and rationals which is given the interpretation:

$$\bar{y}, m, n, p, q, a_1, \ldots, a_n, b_o, \ldots, b_m, c_1, \ldots, c_q, d_o, \ldots, d_p, w_1, \ldots, w_N.$$

The data for such a program is another list, with the interpretation:

$$i, u_{i-q-m}, \ldots, u_i, y_{i-q-n}, \ldots, y_{i-1}.$$

Since the algorithm (4.7) requires an $(i - \max(m+q, n+q))$th value of either u or y (assuming that $p < q$), the program outputs w_i if $i \leqslant \max(m+q, n+q)$.

Models III and VI of the gas-furnace data do not have the structure (4.7), and so cannot be written in ELML. However, models I, II, IV and V can. The ELML programs which constitute these models are:

Model I - Trivial

   0,0,0,0,0,0,1,+53.8,+53.6,...,+57.0 .

Model II - Mean

   53.5,0,0,0,0,0,1,+.3,+.1,...,+3.5.

Model IV - Deterministic

    22.4,5,2,O,O,.57,.01,0,0,0,-.53,-.37,-.51,1,
    +53.8,+53.6,...,+1.6.

Model V - Stochastic

    2.2,5,2,O,2,.57,.01,0,0,0,-.53,-.37,-.51,.53,.63,1,
    +53.8,+53.6,...,+.2.

The table look-ups of these models are the same as they were for the Algol W versions of section 3.7. Table II shows a comparison of the performance of the models.

MODEL	SIZE	INFORMATION GAIN	INFORMATION EXPLAINED
I	1494	O	O
II	1119	375	25.1%
IV	879	615	41.2%
V	853	641	42.9%

TABLE II - Results of Model Assessment Using ELML.

It is seen from the table that model V is now assessed as being better than model IV. However, the assessment of section 3.7, using Algol W, showed model IV to be better than model V (cf. table I).

It is easy to see intuitively why this change in assessment comes about. The language ELML has so much information inherent in its definition, that it requires a comparatively small increase in the number of arbitrary elements of a program to change from the algorithm (of the form (4.7)) appropriate to model IV to that appropriate to model V. Consequently a comparatively small improvement

in the ability of this algorithm to explain the observations
is sufficient to justify the increase.   When the models are
written in Algol W a greater improvement is required.

4.4 Summary

The choice of the programming language which appears in
the characterisation of modelling developed in chapter 3
can be regarded as a specification of the modeller's _a priori_
knowledge about the system which he is investigating.

The characterisation has been extended to deal with the
common case, where increasing sets of observations are being
examined.   It has been shown that in this case the modeller's
a priori beliefs become increasingly irrelevant to the
assessment of any alternative models which he may be considering,
as the observation sets grow.

However, it has been demonstrated that in practice,
especially for small sets of observations, the results of
such an assessment are conditional on the modeller's _a priori_
beliefs.

5.   FRAGMENTS OF PROGRAMMING LANGUAGES

5.1   Introduction

In chapter 4 the idea was developed that the definition
of the programming language to be used for model assessment
corresponds to a specification of the modeller's a priori
knowledge of the system.   But can specific meaning be
attached to the phrase "definition of a programming language"?

When two models are compared, the languages required
to write each of them as programs are rarely exactly the same.
Can their comparison on the basis of program lengths be
meaningful then, since the a priori knowledge assumed for each
is slightly different?

This chapter and the next are concerned with these
questions.   The expedient adopted earlier, of associating
a programming language with a partial recursive function, is
no longer satisfactory, since it gives no information about
the algorithm used for computing the function.

Appendix A reviews one method of defining programming
languages formally, namely the so-called "Vienna Method".
We henceforth assume familiarity with the contents of this
Appendix.   In this chapter, the Vienna method is used as
a basis for a precise definition of "programming language".
The notions of a program being "written in a language" and
of "the function computed by a language" are then introduced.
The concept of "fragment of a language" is made precise in
section 5.4, and is used to define "equivalence of languages".

Languages are equivalent only if they are indistinguishable
to the user.   However, equivalent languages need not be
equal, since their interpreting automata, for instance,
may be different.

Most programs are written in infinitely many languages,
many of which are fragments of others.   Therefore the
"family of languages in which a program is written" is
introduced.   Such a family corresponds, roughly, to the set
of languages which is referred to by a name like "Algol" or
"Fortran".   If suitably restricted, a family of languages
in which a program is written has a "smallest" element, which
is called the "support" of the program.   The main aim of
this chapter is the formalisation of this concept of "the
support of a program", since this is the concept which is
required for model assessment.

The following example may help to clarify this chapter.
Consider the program:

```
    begin

        integer  i,j;

        if i=j then i:=j+1 else i:=j;

        i:=1;

    end.
```

This program is written in Algol, but it uses only a few
of the features which Algol usually provides.   It uses
block structure, some integer arithmetic, assignment statements,
and conditional expressions.   It does not use procedures,
for statements, or goto statements, for instance.   Also it

uses very few terminals.   In our terminology this program
is written in an Algol-family of languages (more precisely,
it is written in an Algol X-family of languages, where Algol X
is some completely specified Algol-like language).   Languages
in this family include all the versions of Algol found
implemented in practice.   The Algol X-support of the program
is, roughly speaking, the smallest Algol-like fragment which
allows the above program to be written.   It will include
block structure, integer addition, assignment statements,
conditional expressions, the identifiers i,j, and the integer
1.   It will not include any further features.   A constraint
on the concrete syntax of the Algol X-family ensures that
the Algol X-support of the program allows more than one
program to be written in it.   Thus the Algol X-support
may not have the trivial and useless concrete syntax:
<program>::=<u>begin</u> <u>integer</u> i,j; <u>if</u> i =j <u>then</u> i:=j+1 <u>else</u> i:=j; i:=1;<u>end</u>
(However, a λ-support of the program exists for some language λ,
which does have this concrete syntax).

   The notion of "equivalence" which will be formalised
is intended to coincide with the concept which appears in
Ollongren (49), although the notion of "fragment" is slightly
different.

   The application of these ideas to model assessment
will be presented in chapter 6.

5.2   Preliminaries

   We shall use p to denote a program in the usual sense,
namely a finite string of letters from a finite alphabet

of terminals.   When discussing a particular language with
a specified (concrete) grammar G, we shall assume the existence
of a parsing algorithm π_G, and write p_G to denote the derivation
tree $\pi_G(p)$ (see section A.3 and Ollongren (49) sections 2.3
and 3).   We shall always assume that G is unambiguous,
namely that for any p, there is at most one derivation of p
in G.   L(G) will denote the language generated by G.

Definition (5.2.1)

Consider three context-free grammars (cf. sec. A.3)

$$G_i = (N_i, \Sigma_i, P_i S_i) \quad , \quad i=1,2,3,$$

where N_i is a set of nonterminals, Σ_i is a set of terminals,
P_i is a set of productions, and S_i is a start symbol.

(i) $G_1 \subseteq G_2 \Longleftrightarrow$ (a) $N_1 \subseteq N_2$

(b) $\Sigma_1 \subseteq \Sigma_2$

(c) $P_1 \subseteq P_2$

(d) $S_1 = S_2$

(ii) $G_1 = G_2 \cap G_3 \Longleftrightarrow$ (a) $N_1 = N_2 \cap N_3$

(b) $\Sigma_1 = \Sigma_2 \cap \Sigma_3$

(c) $P_1 = P_2 \cap P_3$

(d) $S_1 = S_2 = S_3$.

(Note: we use $\subseteq$ to denote improprer inclusion.   Thus
$G \subseteq G$, for all G).

Lemma (5.2.2) $G_1 \subseteq G_2$ & $G_2 \subseteq G_3 \Rightarrow G_1 \subseteq G_3$

Proof: Immediate from transitivity of set inclusion.

__Lemma (5.2.3)__   (i) $G_1 \subseteq G_2 \Rightarrow L(G_1) \subseteq L(G_2)$

(ii) $G_1 = G_2 \cap G_3 \Rightarrow L(G_1) \subseteq L(G_2) \cap L(G_3)$.

__Proof:__   (i) $G_1 \subseteq G_2 \,\&\, p\varepsilon L(G_1) \Rightarrow G_1 \subseteq G_2 \,\&\, S \underset{G_1}{\overset{*}{\Rightarrow}} p$

$$\Rightarrow S \underset{G_2}{\overset{*}{\Rightarrow}} p \quad \text{(by def. (5.2.1))}$$

$$\Rightarrow p \;\varepsilon\; L(G_2)$$

(ii) From (i), $G_1 = G_2 \cap G_3 \Rightarrow L(G_1) \subseteq L(G_2) \,\&\, L(G_1) \subseteq L(G_3)$

$$\Rightarrow L(G_1) \subseteq L(G_2) \cap L(G_3).$$

(Note that the converse inclusion does not hold, because $S \underset{G_2}{\overset{*}{\Rightarrow}} p$ and $S \underset{G_3}{\overset{*}{\Rightarrow}} p$ may be different derivations, neither of which is possible in G_1).

__Definition (5.2.4)__ $\Pi(G) = \{\pi_G(p) : p\varepsilon L(G)\}$ is the set of derivation trees generated by G.

__Lemma (5.2.5)__   (i) $G_1 \subseteq G_2 \Rightarrow \Pi(G_1) \subseteq \Pi(G_2)$

(ii) $G_1 = G_2 \cap G_3 \Rightarrow \Pi(G_1) \subseteq \Pi(G_2) \cap \Pi(G_3)$.

__Proof:__   (i)   From definition (5.2.1) it follows that every production in G_1 is also in G_2.   Hence every derivation in G_1 is also in G_2.   But to every derivation there corresponds a unique derivation tree (Ollongren (49) sec. 2.3).   Hence every derivation tree in $\Pi(G_1)$ is also in $\Pi(G_2)$.

(ii)   From (i), $G_1 = G_2 \cap G_3 \Rightarrow \Pi(G_1) \subseteq \Pi(G_2) \,\&\, \Pi(G_1) \subseteq \Pi(G_3)$

$$\Rightarrow \Pi(G_1) \subseteq \Pi(G_2) \cap \Pi(G_3).$$

The availability of a metalanguage for the definition

of programming languages is assumed, as is the existence
of the set of tree-structured objects which is introduced
in Appendix A.   In the context of model assessment, the
metalanguage is the same as the one that would be used
by the modeller for specifying (informally) his _a priori_
knowledge.

The following definition of an interpreting automaton
differs slightly from that given in section A.6.   Explicit
reference to the set of tree-structured objects has been
dropped, since the same set is assumed throughout.   Also,
a set of error states E is introduced.   Its purpose will
be seen later.

Defintion (5.2.6)

An _interpreting automaton_ is a 5-tuple

$$M = (\text{is-state}, \xi_o, \Lambda, F, E)$$

where is-state is a predicate over the set of objects, ξ_o
is the initial state, Λ is the state-transition function,
$F \subseteq \text{is-state}$ is a set of final states, and E is a set of
error states which has the property:

$$\xi \in E \,\&\, \xi \notin F \Rightarrow \Lambda(\xi) \subseteq E.$$

The predicate is-state is not further defined, but it is
assumed that every state has a (possibly empty) control part,
and that this control determines the state-transition function
in the manner described for LML in section A.6.2, and for
more general languages by Ollongren (49).   (Thus it is
assumed that $\Lambda(\xi)$ is a _set_ of states, in general).   If
$\xi \in F$, then $\Lambda(\xi)$ is not defined.

Definition (5.2.7)

A <u>computation</u> of an interpreting automaton M=(is-state, ξ_o,Λ,F,E) is a sequence $(\xi_o,\xi_1,\ldots)$, such that is-state (ξ_i) for i=0,1,..., and $\xi_i \notin F \Rightarrow \xi_{i+1} \in \Lambda(\xi_i)$. The computation <u>terminates</u> if $\xi_i \in F$ for some i.

5.3 Programming Languages

Definition (5.3.1)

A <u>programming language</u> is a 5-tuple $\lambda =$(G, is-program,T,M,op) where G is an unambiguous, context-free grammar which defines the concrete syntax; is-program is a predicate which defines the abstract syntax; $T:\Pi(G) \to is\hat{-}program$ is a translator; M is an interpreting automaton whose state-transition function is effectively computable, and whose initial state $\xi_o(p_A)$ depends on an object $p_A=T(p_G) \in is\hat{-}program$; and op is an output function whose domain is in the set F of final states of M. Furthermore, for every $p_A \in is\hat{-}program$, if $(\xi_o(p_A),$ $\xi_1^i(p_A), \ldots,\xi_F^i(p_A))$ is a terminating computation, then every other computation $(\xi_o(p_A),\xi_1^j(p_A),\ldots)$ terminates (with $\xi_q^j \in F)$, and $op(\xi_q^j(p))=op(\xi_F^i(p))$.

In the above definition it is assumed that the predicate is-program specifies the abstract syntax not only of the "program", but also of the "data". This constitutes a departure from the practice followed in the Vienna method, and in the definition of LML in Appendix A. To see that this is no restriction, consider the following. If a predicate is-program has been defined which does not make provision for data, and the data is assumed to be located in some directory of the state (as is done in the case of LML), then a new predicate can be defined by:

is-program$_1$ =(<s-program:is-program>, <s-data:is-data>)

where is-data is a predicate satisfying the abstract syntax of data sets. All that remains to be done is to modify the instruction associated with the initial state, so that the first action of the interpreting automaton is to read s-data (p) into the appropriate directory (or directories), and to make s-program (p) the next argument. ξ_1 of the new machine is then identical with ξ_0 of the old one.

The output function op is introduced in order to have available the notion of "result of a computation" without restricting the predicate is-state. Note that although the states appearing in alternative computations need not be the same, we impose the usual requirement that if the result of a computation is defined, then it is unique.

Definition (5.3.2)

Let λ=(G, is-program, T,M,op) be a programming language. p is <u>written in</u> λ (p$\epsilon_w\lambda$) if and only if pϵL(G), p_A=T(p_G) is defined (where p_G=π_G(p)), and for every ξ_i appearing in the computation $(\xi_0(p_A),\xi_1(p_A),\ldots)$, $\Lambda(\xi_i)$ is defined unless $\xi_i\epsilon$F, and $\xi_i\notin$E.

The role of the set of error states E can now be seen. At the end of section A.6 it is remarked that specifying the concrete and abstract syntax of a language is not sufficient to define the valid programs in a language. Part of the definition must be accomplished by specification of the interpreting functions: whenever an invalid program is encountered, an error state is entered. This aspect of language definition is formalised in definition (5.3.2). By definition (5.2.6), once a computation enters an error state it remains in an error state.

Definition (5.3.3)

$P_\lambda = \{p : p \epsilon_w \lambda\}$ is the set of all programs written in λ.

Definition (5.3.4)

The __function computed by the language__ λ is the function $\phi_\lambda : P_\lambda \rightarrow$ range (op), such that $\phi_\lambda(p) = op(\xi_F(p_A))$, where $p_A = T(\pi_G(p))$, $(\xi_0(p_A), \ldots, \xi_F(p_A))$ is a computation and $\xi_F(p_A) \epsilon F$, and $\phi_\lambda(p)$ is undefined if no such ξ_F exists. (Definition (5.3.1) ensures that ϕ_λ is a function).

The function ϕ_λ is a partial effectively computable function. If the sets P_λ and range (op) are suitably arithmetised (i.e. put into one-to-one correspondence with the nonnegative integers), then, by Church's thesis, ϕ_λ can be regarded as a partial recursive function. Furthermore, if the predicate is-program allows a clear distinction between "program" and "data", and if each of these can be arithmetised separately, then φ_λ can be regarded as a partial recursive function of two arguments. If φ_λ is universal, when so regarded (cf. Rogers (9)), then λ is universal. All those entities conventionally considered to be programming languages are universal. However LML, which is included in definition (5.3.1), is not

5.4   Fragments of Languages

In this section a subscript usually denotes a programming language to which reference is being made. For example, G_λ denotes the grammar of λ.

Definition (5.4.1)

Let λ_1 and λ_2 be programming languages. λ_1 is **a fragment** of λ_2 ($\lambda_1 <_F \lambda_2$) if and only if

(i) $G_{\lambda_1} \subseteq G_{\lambda_2}$.

(ii) $p\varepsilon_w\lambda_1 \Longrightarrow p\varepsilon_w\lambda_2$

(iii) $\forall \, p\varepsilon_w\lambda_1$, $\varphi_{\lambda_1}(p) = \varphi_{\lambda_2}(p)$, where it is understood that if one of $\varphi_{\lambda_1}(p)$, $\varphi_{\lambda_2}(p)$ is undefined, then so is the other,

(iv) $\forall \, p\varepsilon L(G_{\lambda_1})$, $p\varepsilon_w\lambda_2 \Longrightarrow p\varepsilon_w\lambda_1$.

Roughly speaking, if λ_1 is a fragment of λ_2, then everything that can be done using λ_1 can also be done using λ_2. However, to make this definition useful, the languages concerned must be defined in a rather idiosyncratic manner. Part (iv) of the definition implies that if λ_2 contains

standard procedures, for example, which are not available
in λ_1, then they must be called by new terminal characters
which do not appear in the grammar of λ_1. This is not the
usual practice, but there is no reason why it should not be
done. A given programming language (understood informally)
does not have a unique formal definition. Definition
(5.4.1) assumes that much more of the burden of the language
definition has been transferred from the translator to the
concrete syntax, than is convenient in practice. This
point will be discussed further in section 6.2.

Theorem (5.4.2)

$<_F$ is reflexive and transitive.

Proof: Reflexivity is obvious

Suppose $\lambda_1 <_F \lambda_2$ and $\lambda_2 <_F \lambda_3$.

Then   (i) $G_{\lambda_1} \subseteq G_{\lambda_2}$ and $G_{\lambda_2} \subseteq G_{\lambda_3}$. Hence $G_{\lambda_1} \subseteq G_{\lambda_3}$ by lemma (5,2,2)

(ii) $p\varepsilon_w\lambda_1 \Rightarrow p\varepsilon_w\lambda_2$ and $p\varepsilon_w\lambda_2 \Rightarrow p\varepsilon_w\lambda_3$. Hence

$p\varepsilon_w\lambda_1 \Rightarrow p\varepsilon_w\lambda_3$.

(iii) $\forall p\varepsilon_w\lambda_1$, $\varphi_{\lambda_1}(p) = \varphi_{\lambda_2}(p)$ and $\forall p\varepsilon_w\lambda_2$,

$\varphi_{\lambda_2}(p) = \varphi_{\lambda_3}(p)$. So $\forall p\varepsilon_w\lambda_1$, $\varphi_{\lambda_1}(p) = \varphi_{\lambda_3}(p)$, by def. (5.4.1) (ii).

(iv) $\forall p\varepsilon L(G_{\lambda_1})$, $p\varepsilon_w\lambda_2 \Rightarrow p\varepsilon_w\lambda_1$ (A)

and $\forall p\varepsilon L(G_{\lambda_2})$, $p\varepsilon_w\lambda_3 \Rightarrow p\varepsilon_w\lambda_2$.   (B)

Suppose $p\varepsilon L(G_{\lambda_1})$ and $p\varepsilon_w\lambda_3$. Then $p\varepsilon L(G_{\lambda_2})$ by lemma

(5.2.3) (i). So $p\varepsilon_w\lambda_2$, by (B). Hence $p\varepsilon_w\lambda_1$, by (A).

Definition (5.4.3)

λ_1 is __equivalent__ to λ_2 $\quad (\lambda_1 \sim \lambda_2)$ if and only if $\lambda_1 <_F \lambda_2$ and $\lambda_2 <_F \lambda_1$.

Theorem (5.4.4)　$\sim$ is an equivalence relation.

Proof: Reflexivity and symmetry are obvious.　Transitivity follows from theorem (5.4.2).

Theorem (5.4.5)　$\lambda_1 \sim \lambda_2 \Longrightarrow$ (i)　$P_{\lambda_1} = P_{\lambda_2}$

$$\text{(ii)} \quad \varphi_{\lambda_1} = \varphi_{\lambda_2}$$

Proof: Immediate from definitions (5.4.1) and (5.4.3).

Lemma (5.4.6)　$\mu <_F \lambda \,\&\, \nu <_F \lambda \,\&\, G_\mu = G_\nu \Longrightarrow \mu \sim \nu$

Proof:　$\mu <_F \lambda \Longrightarrow$ (i)　$p \varepsilon_w \mu \Longrightarrow p \varepsilon_w \lambda$

$\qquad\qquad$ (ii) $\forall p \varepsilon_w \mu,\ \varphi_\mu(p) = \varphi_\lambda(p)$

$\qquad\qquad$ (iii) $\forall p \varepsilon L(G_\mu),\ p \varepsilon_w \lambda \Longrightarrow p \varepsilon_w \mu.$

$G_\mu = G_\nu \,\&\, \nu <_F \lambda \Longrightarrow$ (i)　$p \varepsilon_w \nu \Longrightarrow p \varepsilon_w \lambda$

$\qquad\qquad$ (ii) $\forall\ p \varepsilon_w \nu, \varphi_\nu(p) = \varphi_\lambda(p)$

$\qquad\qquad$ (iii) $\forall\ p \varepsilon L(G_\mu), p \varepsilon_w \lambda \Longrightarrow p \varepsilon_w \nu.$

$\therefore \mu <_F \nu \,\&\, \nu <_F \lambda \,\&\, G_\mu = G_\nu \Longrightarrow$

$\qquad\qquad$ (i)　$G_\mu = G_\nu$

$\qquad\qquad$ (ii) $p \varepsilon_w \mu \Longleftrightarrow p \varepsilon_w \nu$

$\qquad\qquad$ (iii) $\forall\ p \varepsilon_w \mu,\ \varphi_\mu(p) = \varphi_\nu(p)$

$\qquad\qquad$ (iv) $\forall\ p \varepsilon L(G_\mu),\ p \varepsilon_w \nu \Longleftrightarrow p \varepsilon_w \mu.$

Hence $\mu \sim \nu$ by defs (5.4.1) and (5.4.3).

Definition (5.4.7)

$\qquad [\lambda]_\sim = \{\lambda' : \lambda' \sim \lambda\}$ is the equivalence class of λ modulo $\sim$.

Lemma (5.4.8) For any programming language λ, there exists at most a finite number of equivalence classes $[\mu]_\sim$, such that $\mu <_F \lambda$.

Proof: $\mu <_F \lambda \Rightarrow G_\mu \subset G_\lambda = (N_\lambda, \Sigma_\lambda, P_\lambda, S_\lambda)$ (cf. def. (5.2.1)). But each of the sets $N_\lambda, \Sigma_\lambda, P_\lambda, S_\lambda$ is finite.   Hence, by definition (5.2.1), there exist only finitely many distinct G_μ, such that $G_\mu \subset G_\lambda$.   The result follows from lemma (5.4.6).

Definition (5.4.9)

Let S be a set of programming languages.   Then the set of common fragments of S is the set

$$\bigcap_F S = \{\lambda: \mu \varepsilon S \Rightarrow \lambda <_F \mu\}$$

Theorem (5.4.10) $\lambda \varepsilon \bigcap_F S \& \lambda \sim \mu \Rightarrow \mu \varepsilon \bigcap_F S.$

Proof: $\lambda \varepsilon \bigcap_F S \& \lambda \sim \mu \Rightarrow (\nu \varepsilon S \Rightarrow \lambda <_F \nu) \& \lambda \sim \mu$

$\Rightarrow (\nu \varepsilon S \Rightarrow \lambda <_F \nu) \& \mu <_F \lambda$

$\Rightarrow (\nu \varepsilon S \Rightarrow \mu <_F \nu)$ by theorem (5.4.2)

$\Rightarrow \mu \varepsilon \bigcap_F S$ by definition (5.4.9).

Definition (5.4.11)

The greatest common fragment of a set of programming languages S is the set

$$\Gamma(S) = \{\lambda : \lambda \varepsilon \bigcap_F S \& (\mu \varepsilon \bigcap_F S \Rightarrow \mu <_F \lambda)\}.$$

Theorem (5.4.12) $\lambda \varepsilon \Gamma(S)$ & $\mu \varepsilon S \Rightarrow \lambda <_F \mu.$

Proof: $\lambda \varepsilon \Gamma(S) \Rightarrow \lambda \varepsilon \bigcap_F S \Rightarrow \lambda <_F \mu.$

Theorem (5.4.13)   (i) $\lambda \varepsilon \Gamma(S) \& \mu \varepsilon \Gamma(S) \Rightarrow \lambda \sim \mu$

(ii) $\lambda \varepsilon \Gamma(S) \& \lambda \sim \mu \Rightarrow \mu \varepsilon \Gamma(S)$

<u>Proof</u>: (i) $\lambda\epsilon\Gamma(S)\&\mu\epsilon\Gamma(S)\Longrightarrow\lambda\epsilon\Gamma(S)\&\mu\epsilon\bigcap_F S$

$$\Longrightarrow\mu<_F\lambda, \text{ by definition (5.4.11)}.$$

Similarly, $\lambda<_F\mu$.   Hence $\lambda\sim\mu$.

(ii) $\lambda\epsilon\Gamma(S)\&\lambda\sim\mu\Longrightarrow\lambda\epsilon\bigcap_F S\&(\nu\epsilon\bigcap_F S\Longrightarrow\nu<_F\lambda)\&\lambda\sim\mu$

by definition (5.4.11),

$$\Longrightarrow\mu\epsilon\bigcap_F S\&(\nu\epsilon\bigcap_F S\Longrightarrow\nu<_F\lambda)\&\lambda<_F\mu$$

by theorem (5.4.10) and definition (5.4.3),

$$\Longrightarrow\mu\epsilon\bigcap_F S\&(\nu\epsilon\bigcap_F S\Longrightarrow\nu<_F\mu)$$

by theorem (5.4.2),

$$\Longrightarrow\mu\epsilon\Gamma(S), \text{ by definition (5.4.11)}.$$

<u>Lemma (5.4.14)</u> $\lambda\epsilon S\&\lambda\epsilon\bigcap_F S\Longrightarrow\lambda\epsilon\Gamma(S)$

<u>Proof</u>: $\lambda\epsilon S\&\lambda\epsilon\bigcap_F S\Longrightarrow(\mu\epsilon\bigcap_F S\Longrightarrow\mu<_F\lambda)\&\lambda\epsilon\bigcap_F S$

$$\Longrightarrow\lambda\epsilon\Gamma(S), \text{ by definition (5.4.11)}.$$

Definition (5.4.15)

If $p\epsilon_w\lambda$, then the <u>λ-family of languages</u> in which p is written is the set

$$\Delta_\lambda(p)=\{\mu:p\epsilon_w\mu\&(\lambda<_F\mu \text{ or } \mu<_F\lambda)\}.$$

If $p\not\epsilon_w\lambda$, then $\Delta_\lambda(p)=\emptyset$.

<u>Theorem (5.4.16)</u> $p\epsilon_w\lambda\Longleftrightarrow\lambda\epsilon\Delta_\lambda(p)$

<u>Proof</u>: Obvious

<u>Theorem (5.4.17)</u> $\mu\epsilon\Delta_\lambda(p)\&\nu\epsilon\Delta_\lambda(p)\Longrightarrow\varphi_\mu(p)=\varphi_\nu(p)$.

<u>Proof</u>: $\mu\epsilon\Delta_\lambda(p)\Longrightarrow(\mu<_F\lambda \text{ or } \lambda<_F\mu)\&p\epsilon_w\mu\&p\epsilon_w\lambda$.

$\mu<_F\lambda\&p\epsilon_w\mu\Longrightarrow\varphi_\lambda(p)=\varphi_\mu(p)$ by definition (5.4.1)

and $\lambda<_F\mu\&p\epsilon_w\lambda\Longrightarrow\varphi_\mu(p)=\varphi_\lambda(p)$ similarly.

Hence $\mu\epsilon\Delta_\lambda(p)\Longrightarrow\varphi_\mu(p)=\varphi_\lambda(p)$.

Similarly $\nu\epsilon\Delta_\lambda(p)\Longrightarrow\varphi_\nu(p)=\varphi_\lambda(p)$.

Therefore $\mu\epsilon\Delta_\lambda(p)\,\&\,\nu\epsilon\Delta_\lambda(p)\Longrightarrow\varphi_\mu(p)=\varphi_\nu(p)$.

Note that if the convention $p\not\epsilon_w\lambda\Longrightarrow\Delta_\lambda(p)=\emptyset$ were dropped, then theorem (5.4.17) would not hold. Also, note that if $p\epsilon_w\lambda$ then it is always possible to construct a μ, such that $p\epsilon_w\mu$, yet $\Delta_\lambda(p)\cap\Delta_\mu(p)=\emptyset$.

Definition (5.4.18)

The $\underline{\lambda\text{-support}}$ of a program p is the set of languages
$$\Sigma_\lambda(p)=\Gamma(\Delta_\lambda(p)).$$

By theorem (5.4.13) (i), all languages in the λ-support of p are equivalent. If $p\not\epsilon_w\lambda$, then $\Sigma_\lambda(p)=\emptyset$.

We now come to the central result of this chapter.

Theorem (5.4.19)

(i) $\forall\;\lambda',\lambda''\epsilon\Delta_\lambda(p)$, $\exists\;\mu\epsilon\Delta_\lambda(p)$ such that
$\mu<_F\lambda'$ and $\mu<_F\lambda''$.

(ii) $\Delta_\lambda(p)\neq\emptyset\Longrightarrow\exists\nu\epsilon\Delta_\lambda(p)$, such that $\forall\lambda'\epsilon\Delta_\lambda(p)$,

$\nu<_F\lambda'$ (i.e, there exists a "smallest" language in $\Delta_\lambda(p)$).

(iii) Furthermore, $\nu\epsilon\Sigma_\lambda(p)$.

Proof: (i) Suppose $\lambda'\epsilon\Delta_\lambda(p)$ and $\lambda''\epsilon\Delta_\lambda(p)$.

Then $p\epsilon_w\lambda'\,\&\,(\lambda<_F\lambda'\text{ or }\lambda'<_F\lambda)\,\&\,p\epsilon_w\lambda''\,\&\,(\lambda<_F\lambda''\text{ or }\lambda''<_F\lambda)$.

If $\lambda'<_F\lambda\,\&\,\lambda<_F\lambda''$ then $\lambda'<_F\lambda''$ by theorem (5.4.2). Put $\mu=\lambda'$.

If $\lambda'' <_F \lambda \,\&\, \lambda <_F \lambda'$ then $\lambda'' <_F \lambda'$, similarly. Put $\mu = \lambda''$.

If $\lambda <_F \lambda'$ & $\lambda <_F \lambda''$, put $\mu = \lambda$.

If $\lambda' <_F \lambda$ & $\lambda'' <_F \lambda$, then μ can be constructed as follows.

Take $G_\mu = G_{\lambda'} \cap G_{\lambda''}$

$\quad$ is-$\widehat{\text{program}}_\mu = \{ T_{\lambda'}(p_{G_{\lambda'}}) : p_{G_{\lambda'}}' \in \Pi(G_\mu) \}$

$\quad T_\mu(p_{G_{\lambda'}}') = T_{\lambda'}(p_{G_{\lambda'}}'), \forall p_{G_{\lambda'}}' \in \Pi(G_\mu)$

$\quad M_\mu = M_{\lambda'}$

$\quad op_\mu = op_{\lambda'}$.

Now $p \in L(G_{\lambda'})$ & $p \in L(G_{\lambda''})$ & $G_{\lambda'} \subseteq G_\lambda$ & $G_{\lambda''} \subseteq G_\lambda$. If the

derivations $S \underset{G_{\lambda'}}{\overset{*}{\Longrightarrow}} p$ and $S \underset{G_{\lambda''}}{\overset{*}{\Longrightarrow}} p$ were distinct then there would

be two distinct derivations $S \underset{G_\lambda}{\overset{*}{\Longrightarrow}} p$, by definition (5.2.1).

But this would contradict the standing assumption that G_λ

is unambiguous. Consequently the derivations of p in $G_{\lambda'}$

and $G_{\lambda''}$ are the same. Consequently the productions which

appear in this derivation are all in G_μ. Hence $p \in L(G_\mu)$,

and $p_{G_\mu} = \pi_{G_\mu}(p) \in \Pi(G_\mu)$. Therefore $T_\mu(p_{G_\mu}) = T_{\lambda'}(p_{G_\mu}) \in$ is-$\widehat{\text{program}}_\mu$.

$\quad$ Also, $p \varepsilon_w \lambda' \Rightarrow \Lambda_\mu(\xi_i)$ is defined for every ξ_1 in the

computation $(\xi_0(p), \xi_1(p), \ldots)$ $(\xi_i \notin F_\mu)$, and $\Lambda_\mu(\xi_i) \cap E = \emptyset$.

These last two properties follow becaue the interpreting

automata of λ' and μ are the same. Hence, by definition

(5.3.2), $p \varepsilon_w \mu$ . . . . (1)

From the construction, it is clear that:

(a) $G_\mu \subseteq G_{\lambda'}$

(b) $p' \varepsilon_w \mu \Rightarrow p' \varepsilon_w \lambda'$

(c) $\forall \ p' \varepsilon_w \mu \ : \ \varphi_{\lambda'}(p') = \varphi_\mu(p')$

(d) $\forall \ p' \varepsilon L(G_\mu) \ : \ p' \varepsilon_w \lambda' \Rightarrow p' \varepsilon_w \mu$.

Hence, by definition (5.4.1), $\mu <_F \lambda'$. . . . . (2)

But $\lambda' <_F \lambda$ by supposition, hence $\mu <_F \lambda$ by theorem (5.4.2)..(3)

So, from (1) and (3), $\mu \varepsilon \Delta_\lambda(p)$ . . . . (4)

It remains to show that $\mu <_F \lambda''$:

(e) From the construction, it is clear that $G_\mu \subseteq G_{\lambda''}$.

(f) $p' \varepsilon_w \mu \Rightarrow p' \varepsilon_w \lambda$, since $\mu <_F \lambda$.   But $p' \varepsilon L(G_{\lambda''})$

by (e) and lemma (5.2.3) (i), and $\lambda'' <_F \lambda$ by supposition.
Hence $p' \varepsilon_w \lambda''$, by definition (5.4.1) (iv).

(g) $p' \varepsilon_w \mu \ \& \ \mu <_F \lambda \Rightarrow \varphi_\mu(p') = \varphi_\lambda(p')$ by definition (5.4.1) (iii).

$p' \varepsilon_w \lambda'' \ \& \ \lambda'' <_F \lambda \Rightarrow \varphi_{\lambda''}(p') = \varphi_\lambda(p')$ similarly.

So, using (f), $p' \varepsilon_w \mu \Rightarrow \varphi_\mu(p') = \varphi_{\lambda''}(p')$.

(h) Let $p' \varepsilon L(G_\mu)$ and $p' \varepsilon_w \lambda''$.   Then $p' \varepsilon_w \lambda$, since

$\lambda'' <_F \lambda$.   But $\mu <_F \lambda$, so $p' \varepsilon_w \mu$, by definition (5.4.1) (iv).

(e), (f), (g), (h) together show that $\mu <_F \lambda''$ . . . . . . (5)

(2), (4) and (5) together prove (i).

(ii) By lemma (5.4.8), there are at most finitely

many $[\lambda']_{\sim} \subseteq \Delta_\lambda(p)$, such that $\lambda' <_F \lambda$.   This, together with (i),

shows that $\exists \nu \varepsilon \Delta_\lambda(p)$, such that $\forall \lambda' \varepsilon \Delta_\lambda(p)$, $\nu <_F \lambda'$.

(iii) By (ii), $\nu \varepsilon \cap_F \Delta_\lambda(p)$ and $\nu \varepsilon \Delta_\lambda(p)$.   Hence, by lemma

(5.4.14), $\nu\epsilon\Sigma_\lambda(p)$.

Corollary (5.4.20) $p\epsilon_w\lambda \Longrightarrow \Sigma_\lambda(p) \subset \Delta_\lambda(p)$

Definition (5.4.21)   Let S and T be sets of programming languages.   Then $S\smallsmile T \Longleftrightarrow (\forall s\epsilon S \& \forall t\epsilon T): s\smallsmile t$.

Theorem (5.4.22) $\mu\epsilon\Sigma_\lambda(p_1)$ & $\nu\epsilon\Sigma_\lambda(p_2)$ & $\mu\smallsmile\nu \Longrightarrow \Sigma_\lambda(p_1)\smallsmile\Sigma_\lambda(p_2)$.

Proof: Immediate from theorems (5.4.13) (i) and (5.4.4).

Definition (5.4.23)   Let $\tau(p)$ denote the set of terminals which occur in p.

Theorem (5.4.24) $\Sigma_\lambda(p_1)\smallsmile\Sigma_\lambda(p_2) \Longrightarrow \tau(p_1)=\tau(p_2)$.

Proof: $\Sigma_\lambda(p_1)\smallsmile\Sigma_\lambda(p_2) \Longrightarrow G_{\Sigma_\lambda(p_1)}=G_{\Sigma_\lambda(p_2)}$, by definition

(5.4.3).   Let $G_{\Sigma_\lambda(p_1)}=(N,A,P,S)$, where A is the alphabet of terminals.   Suppose $\tau(p_1)\neq\tau(p_2)$.   Then $\tau(p_1)\subset A$ and $\tau(p_2)\subset A$ and either $\tau(p_1)\neq A$ or $\tau(p_2)\neq A$ (or both).   Suppose $\tau(p_1)\neq A$.   Consider the grammar $G=(N,\tau(p_1), P',S)$, where P'

is obtained from P by deleting those production rules $\xi\rightarrow\beta$ for which $\beta\epsilon A-\tau(p_1)$.   Clearly, $G\subset G_{\Sigma_\lambda(p_1)}$.   If $\Sigma_\lambda(p_1)$ is

now replaced by the language obtained from $\Sigma_\lambda(p_1)$ by replacing

$G_{\Sigma_\lambda(p_1)}$ by G, then the resulting language will be a fragment of $\Sigma_\lambda(p_1)$ and will be in $\Delta_\lambda(p_1)$.   But this is a contradiction, since $\Sigma_\lambda(p_1)$ is a fragment of every language in $\Delta_\lambda(p_1)$, and is not equivalent to the new language (because $\tau(p_1)\neq A$).

Hence $\tau(p_1)=A$.   Similarly $\tau(p_2)=A$.   Hence $\tau(p_1)=\tau(p_2)$.

5.5   Conclusion

Conditions (i) and (iv) of definition (5.4.1) ensure
that it is possible for two different programs to have the
same λ-support.   Recall the short program given in section
5.1.   Consider its AlgolW-support, where AlgolW is taken
to be defined (informally) as in (50).   Condition (i) of
definition (5.4.1) implies that its concrete syntax specification
must include the production rules:

<program>::=<block>.

<block> ::=<block body><statement>end

<block body>::=<block head>|<block body><statement>;

<block head>::= begin|<block head><declaration>

<statement>::=<if statement>|<simple statement>

<if statement>::=<if clause><simple statement>

                else <statement>

etc.

The language generated by these production rules includes
the program

begin

    integer i,j;

    if i=j then i:=j else i:=j+1;

    j:= i;

end.

Now, condition (iv) of definition (5.4.1) ensures that this
program is written in the AlgolW-support of the first
program, since it is obviously written in AlgolW.   It is
easy to see that the AlgolW-supports of the two programs are

equivalent. However, a proof of this would require a full formal definition of AlgolW.

In the above example, we have relied on an informal understanding of the semantics of AlgolW, and we were content not to prove the equivalence of the AlgolW-supports of the two programs. It is envisaged that this will be the typical method of using the concept of λ-support in connection with model assessment. The language λ will not need to be specified formally, and theorem (5.4.24) provides an easily-checked necessary condition for the equivalence of λ-supports.

This chapter has shown that if programming languages are considered to be defined in a certain way, (namely, using the Vienna method, and in accordance with the comments on definition (5.4.1)), then it is possible to speak precisely about the"smallest" language required to run a particular program.

The reason for showing this is the following. Suppose that the program is a model, and that the definition of the smallest language required to run it is taken as a specification of the modeller's _a priori_ knowledge. Then it has been shown that this concept of "_a priori_ knowledge" can be precisely defined, if required.

However, we have succeeded only in defining this concept relative to a particular family of languages.

6 λ - COMPARABILITY

6.1 Introduction

Two rival models of a system, written in a language λ,
will very rarely have the same λ-support.   It may be argued,
then, that the use of each of them implies a slightly
different set of a priori assumptions.   In this case, a
comparison of the two models may not seem meaningful.

On the other hand, one may consider that the choice
of a particular language specifies the a priori assumptions,
and that these are not changed if it later appears that certain
features of the language are not needed.

Rather than argue the merits of either viewpoint, we
shall attempt to show that it does not matter much for model
assessment, whichever position is adopted.

For the purposes of model assessment, the term
"programming language" must be understood rather differently
than is usual in computer science.   Since the definition
of Algol 60 it has become common to define "reference"
languages, which are independent of particular hardware
facilities.   Consequently, certain aspects of a language
(such as input/output in Algol) are considered to lie outside
the province of its definition.   For us, however "programming
language" means a computing facility, exactly as it appears
to the programmer.

The most important difference between this view of a
language and a "reference" language (for us) is that standard

procedures, such as input/output and mathmatical procedures
(e.g. sin, sqrt), must be considered to be included in the
language definition.

6.2   λ-Comparable Models

Definition (6.2.1)

Let m_1 and m_2 be two models of a system S.   Then m_1
and m_2 are λ-comparable if and only if

$$\Sigma_\lambda(m_1) \sim \Sigma_\lambda(m_2) \neq \emptyset$$

In this definition it is of course assumed that λ
is a language which interprets m_1 and m_2 as models of S;
in other words, in the terminology of definition (3.3.6),
m_1 and m_2 are assumed to be (λ, E_1) and (λ, E_2) - models of S,
respectively, for some E_1 and E_2.

λ-comparability ensures that all the "facilities" of
a language used by one model are also used by the other.
Such apparently trivial details as whether procedures are
allowed to take one or several arguments, or how many terminal
characters are available, can be regarded as "facilities".
More obviously significant "facilities" are features such
as the availability of goto statements, or of arrays.
λ-comparability ensures the identity of all such features.

Perhaps the feature of a language which can be expected
to affect model assessment most is its complement of standard
procedures - namely, those procedures which can be called
in a program without being defined (declared).   Consequently,

if λ-comparability is to be a useful concept, it must be
ensured that λ-comparable models call the same standard
procedures.   This is so if standard procedure names are
considered to be terminals of the language.   As was mentioned
in chapter 5, this is not the usual practice, but we adopt
it for the following reasons.

Firstly, it is demonstrated in section 6.3 how standard
procedures can be included in the language definition when
their identifiers are treated as terminals.   Secondly, there
is no essential distinction, for our purposes, between a
standard procedure and an operation such as addition,
which **is** usually called by its own terminal.   Such distinction
as there is arises for reasons of convenience of practical
language implementation.   Thirdly, if the syntax of standard
procedure calls were the same as the syntax of programmer-
supplied procedure calls (that is, if standard procedure
identifiers were built up from existing terminals), then our
concept of "fragment of a language", as defined in section
5.4, would not correspond to the intended intuitive notion,
and λ-comparability would not ensure the use of the same
standard procedures.

To see this, suppose that two models of some system
are written in Algol.   Suppose that one of them uses the
standard procedures entier and sqrt, while the other uses
sin as well as entier and sqrt.   If _entier_, _sqrt_ and _sin_
are terminals of the language, then the models are certainly
not Algol-comparable (by theorem (5.4.24)).

But if entier, sqrt and sin are simply procedure
identifiers whose syntax is given by, for example:

    <procedure identifier>::=<identifier>

    <identifier ::= letter>⌈<identifier><letter>

then the Algol-supports of the two models <u>may</u> be
equivalent.   This syntax, which is part of the syntax of
the Algol-support of the first model, allows the identifier
sin to be used (since the productions <letter>::=i|n|s must
also be part of the syntax), and furthermore it allows it
to be used as a procedure identifier.   Now, suppose that the
Algol-support of the first model is a fragment of the
Algol-support of the second model (a possibility which we wish
to allow, intuitively).   Then, according to definition
(5.4.1) (iv), the sin call must have the same effect in both
languages.   So, the Algol-support of the first model must
contain sin as a standard procedure.   Clearly this contradicts
the intended meaning of "Algol-support".

Consequently, we insist that standard procedure identifiers
be regarded as terminals.   If it is now stipulated that only
λ-comparable models should be compared for model assessment,
then we have the formal equivalent of the intuitive idea,
that models should be compared only if they use the same
facilities of a language.   One reason for making this
stipulation has already been referred to in section 6.1.
It may be felt to be an "unfair" comparison if the models
are not λ-comparable.

An obvious example of this would be a comparison of a

difference-equation model with a differential-equation model.
If the differential-equation model were allowed to call a
standard procedure for integration, would it be reasonable
to compare the "number of arbitrary elements" embodied in
it with the number embodied in the difference-equation model?
The difference-equation model requires fewer <u>a priori</u>
assumptions (if its λ- support is a fragment of the λ-support
of the differential-equation model, where λ is the language
in which both models are written).

There are, however, two ways of making models λ-comparable.
Rather than adding an explicitly declared integration procedure
to the differential-equation model, it is possible to add a
"dummy" call of the standard procedure for integration to
the difference-equation model. It is the flexibility
offerred by this possibility, of "padding", models with
redundant statements, that reduces the significance of any
insistence on λ-comparability. This will be demonstrated
in section 6.3.

If λ-comparability is required, there still remains
the choice of a suitable λ-support for the models which are
to be compared. Returning to the above example, there is
still a decision to be made - should both models call the
standard procedure for integration, or neither? This
decision, of course, is very significant for model assessment.
But this is the decision discussed in chapter 4. In other
words, it will be governed by the <u>apriori</u> assumptions that
the modeller wishes to make.

6.3   Example: Algol W-Comparable Gas Furnace Models

This section investigates how the assessment of the six
models of the gas-furnace data (cf. chapter 3) is altered,
if they are required to be AlgolW-comparable.

6.3.1   Standard Procedures

The definition of Algol W is assumed to be a formalised
version of the specification given in (50).   The six models
use three standard procedures, namely the input/output
procedures READ, READON and WRITE.   In accordance with the
discussion of section 6.3, we consider the syntax specification
of (50) to be augmented by the productions:
<simple statement>::=<standard procedure statement>
<standard procedure statement>::=<standard procedure identifier>
                              (<actual parameter list>)
<standard procedure identifier>::=READ| READON|WRITE
The abstract syntax, translator and interpreting
automaton are considered to be modified accordingly.

6.3.2 AlgolW-Comparable Models

In this example the models are modified so as to be
AlgolW-comparable by inserting redundant statments and
expressions, rather than by avoiding certain constructions.
Referring to section 3.7, and comparing the models in
order, we notice first that the syntax of the AlgolW-
support of model II contains the productions
<simple t expression>::=<simple t expression>+<t term>|<t term>

whereas the syntax of the AlgolW-support of model I contains only the production

<simple t expression>::=<t term>

(For the significance of "t" see (50) or the introduction to Appendix B).   This discrepancy can be removed by changing the WRITE statement of model I to: WRITE(Y(I)+O);.

Model III requires several productions which are not needed for models I or II.   These are:

<letter>::= N|U

<unscaled real>::=.<integer number>

<simple t expression>::=<simple t expression>-<t term>

<t term>::=<t term>*<t factor>

<logical expression>::=<relation>

<relation>::=<simple t expression><relational operator>
                                    <simple t expression>

<relational operator>::=<

<statement>::=<if statement>

<simple statement>::=<block>|< t assignment statement>

<t assignment statement>::=<t left part>< t expression>

<t left part>::= <t variable>:=

<if statement>::=<if clause><simple statement>ELSE
                                    <statement>

<if clause>::= IF <logical expression>   THEN

Most but not all of these are needed for model IV, but model IV itself needs two productions which are not needed by models I,II, or III:

<letter>::= E|V|W|Z

<actual parameter list>::=<actual parameter>|<actual parameter list>,
                                              <actual parameter>
The only new production required by

models V and VI are <letter>::=A, and <letter>::=W, respectively,

but these can easily be removed by using different identifiers.

     We give below the six models, modified so as to be

AlgolW-comparable.   The concrete syntax of their common

AlgolW-support is given in Appendix B.

I   The Trivial Model

```
BEGIN
      INTEGER    I,J,N,V,W,Z;  REAL ARRAY E,U,Y(1::296);
      BEGIN
            FOR J:=1 UNTIL 296 DO READON (Y(J-O));
            READ (I);
            V:=O;
            IF I<1 THEN WRITE (O,V) ELSE WRITE(Y(I)*1+O);
      END
END.
53.8      53.6      53.5      ...      57.O
```

II   The Mean

```
BEGIN
      INTEGER    I,J,N,V,W,Z;  REAL ARRAY E,U,Y(1::296);
      BEGIN
            FOR J:=1 UNTIL 296 DO READON (Y(J-O));
            READ (I);
            V:=O;
```

```
                    IF I<1 THEN WRITE (O,V) ELSE
                        WRITE (53.5+Y(I)*1);

        END
END.
.3      .1      O      O      -.1      ...      3.5
```

III   Deterministic Transfer Function (Using Input Observations Only).

```
BEGIN
      INTEGER I,J,E,V,W,Z;   REAL ARRAY N,U,Y(1::296);
      FOR J:=1 UNTIL 296 DO READON (N(J));
      READ (I);
      IF I<6 THEN WRITE (N(I),O) ELSE
      BEGIN
            FOR J:=1 UNTIL 5 DO
            BEGIN
                    Y(J):=N(J)-53.5;
                    READON (U(J));
            END;
            FOR J:=6 UNTIL I DO
            BEGIN
                    READON (U(J));
                    Y(J):=.57*Y(J-1)+.01*Y(J-2)-.53*U(J-3)
                        -.37*U(J-4)-.51*U(J-5);
            END;
            WRITE (Y(I)+53.4+N(I));
      END;
END.
53.8      53.6      ...      4.1
```

IV Deterministic Transfer Function (Using Input and Output

Observations)

```
BEGIN
      INTEGER I,J;   REAL ARRAY E(1::296);
      REAL N,U,V,W,Y,Z;
      FOR J:=1 UNTIL 296 DO READON (E(J));
      READ (I);
      N:=0;
      IF I<6 THEN WRITE (E(I)) ELSE
      BEGIN
            READON (U,V,W,Y,Z);
            WRITE (22.4 -.53*U -.37*V -.51*W + .57*Y +.01*Z+E(I));
      END;
END.
53.8       53.6       ...       1.6
```

V Stochastic Process Model

```
BEGIN
      INTEGER I,J;   REAL ARRAY E,U,Y(1::296);
      REAL N,V,W,Z;
      FOR J:=1 UNTIL 296 DO READON (E(J));
      READ (I);
      IF I<8 THEN WRITE (E(I)) ELSE
      BEGIN
            FOR J:=1 UNTIL 7 DO READON (U(I-J), Y(I-J));
            WRITE (2.1*Y(I-1) - 1.5*Y(I-2) +.34*Y(I-3)
                  +.01*Y(I-4)-.53*U(I-3)+.44*U(I-4)
                  -.28*U(I-5)+.55*U(I-6)-.32*U(I-7)
                  +E(I)+2.2);
```

```
      END;

END.

53.8      53.6      ...      .2
```

VI   Box & Jenkins Model

```
BEGIN

      INTEGER I,J;   REAL ARRAY E,U,Y(1::296);

      REAL N,V,W,Z;

      FOR J:=1 UNTIL 296 DO READON (E(J));

      READ (I);

      If I<8 THEN WRITE (E(I)) ELSE

      BEGIN

            FOR J:=1 UNTIL 7 DO READON (U(I-J),Y(I-J));

            WRITE (2.1*Y(I-1)-1.5*Y(I-2)+.34*Y(I-3)

                  +.01*Y(I-4)-.53*U(I-3)+.44*U(I-4)

                  -.28*U(I-5)+.55*U(I-6)-.32*U(I-7)

                  +E(I)-.57*E(I-1)-.01*E(I-2)+2.2);

      END;

END.

53.8      53.6      ...      .4
```

It will be seen that the models have been "padded" so
that they are not changed in essence, although the details
of their compilation and execution may have been changed
(for example, by the introduction of the extra block in
models I and II).   It is straightforward but tedious to check
that each production listed in Appendix B is used in each of
these models, and that every production required for the

parsing of these models is listed in Appendix B.

6.3.3   Comparison of Assessments

Table III shows a comparison of the sizes of the models
before and after the above modifications.   Table IV compares
their performances.   The table look-ups remain unchanged
(from those shown in Appendix C), since the syntax required
for each of them is already common.

Model	Size, excluding table look-up		Size of table look-up	Total Size	
	Unmodified	Modified		Unmodified	Modified
I	52	93	1480	1532	1573
II	57	96	1102	1159	1198
III	199	209	877	1076	1086
IV	129	131	835	964	966
V	203	212	802	1005	1014
VI	225	234	788	1013	1022

Table III - Sizes of Models Before and After
Modification

Tables III and IV show that although the sizes of models
I and II, excluding the table look-ups, nearly doubled as a
result of the modifications, the information gains and the
information explained by each model were very little affected.
This can be expected to be true in many cases, for the following
reason.   The size of the trivial model will usually be

Model	Information Gain		Information Explained	
	Unmodified	Modified	Unmodified	Modified
I	O	O	O	O
II	373	375	24.4%	23.8%
III	456	487	29.8%	31.0%
IV	568	607	37.1%	38.6%
V	527	559	34.4%	35.4%
VI	519	551	33.8%	35.4%

Table IV - Performances of Models Before and
After Modification

dominated by the size of its table look-up, even for small

observation sets.   Any necessary "padding" of the model can

usually be introduced very economically, so the overall size

of the trivial model does not change much (from 1532 to

1573 in the present example).   On the other hand, as the

models become more sophisticated and their table look-ups

smaller, they will need less "padding" since many more

features of the language will already be in use (cf.table III).

Thus, once again, the overall size will not change much.

6.4   Conclusion

Although chapter 5 provides some precise concepts which

can be associated with "a priori knowledge", it is not clear

which concept is most appropriate.   Should "a priori knowledge"

be considered to be specified by the language λ in which one

intends to write a model, or by the λ-support in which it turns

out to be written?   The first view corresponds to simply
associating "a priori knowledge" with a certain partial
recursive function without reference to the model, while
the second view corresponds to associating it with a partial
recursive function after examining the assumptions needed by
the model.

For the example investigated, it has been shown that
it does not matter, in practice, which view is adopted.
There is reason to suppose that this is true in most cases.
This is most useful, not only because it is not clear whether
models should be λ-comparable before being compared, but
also because checking λ-comparability of complicated models
would be quite difficult (although much of it could be
automated).

7. TABLE LOOK-UP CODINGS

7.1   Introduction

Model assessment will obviously be affected by the manner
in which table look-ups in models are coded.   Since a table
look-up can be viewed as a generalised error (cf. chapter3),
the details of the coding will determine the implicit trade-
off between complexity and approximation.   In the examples
of previous chapters, a particular coding has been assumed
without comment.   We shall show that this coding is a natural
one to use.

7.2   Size-Capturing Codings

We assume a finite alphabet P.   Let P* be the set of
all finite sequences of elements of P, including the empty
sequence Λ.   Suppose that model assessment is to be performed
using a language λ, and that the set of programs written in λ
is a subset of P*.

Suppose further that all the rival models being assessed
are such that the elements of their table look-ups are
rationals.   By using a suitable scaling, all these elements
can be regarded as integers (cf. sec. 3.2).   In order to
obtain a concrete λ-model, these integers must be coded into
elements of P* which appear as fragments of programs written
in λ.

Definition (7.2.1)

A coding is a total, effective, injective function

c:$N \to P^*$, where N denotes the set of all integers (previously N has denoted the set of nonnegative integers).

Let N_+ denote the set of nonnegative integers.

Definition (7.2.2)

With each coding c there can be associated a code-length function $L = c \circ \ell : N \to N_+$, where ℓ is the function which gives the length $\ell(p)$ of $p \epsilon P^*$. Here o denotes composition, so that $L(n) = \ell(c(n))$,

The terms which appear in a table look-up of a model represent the errors which the model makes in trying to compute the observed system behaviour. Our method of assessing models is based on the assumption that a short table look-up corresponds to small errors. This will only be so if the coding used in λ is suitable.

Definition (7.2.3)

A size-capturing coding is one for which the associated code-length L satisfies:

(i) $L(-n) = L(n)$, for all nϵN,

(ii) $|n_1| > |n_2| \Rightarrow L(n_1) \geqslant L(n_2)$.

Clearly, the use of a size-capturing coding leads to table look-ups having the desired property. The requirement that L should be an even function reflects the usual indifference to the sense of the error. The positive entries appearing in Appendix C were considered to be preceded by "+", in order to obtain a size-capturing coding.

Size-capturing codings can be compared with the class L of loss-functions discussed by Deutsch (51). The use of a coding which was not size-capturing would be tantamount to declaring that certain large errors are preferable to certain small errors, or, perhaps, that in some cases positive errors are preferable to negative ones, and so on. This is perfectly permissible, of course, but we wish to restrict ourselves to the more usual situation. Notice also that the same effect could be achieved by using a size-capturing coding, but changing the measurement scales (i.e. by "transforming the variables").

Let Q be a subset of P, containing r elements ($r \geqslant 2$), and C_Q be the set of size-capturing codings whose ranges are subsets of Q^*. For example, the ordinary decimal coding of the integers has as range a subset of $\{0,1...,9,+,-\}^*$.

Definition (7.2.4)

A <u>smallest r-coding</u> is a size-capturing coding $c_0 \epsilon C_Q$, such that $L_0(n) = \left[\log_r |n|\right] + 1$, where $L_0 = c_0 \bullet \ell$ and $\left[x\right]$ denotes the greatest integer not exceeding x ($x \epsilon \mathbb{R}$), and $L_0(0) = 0$.

For example, if $r=2$ then the coding c_0 defined as follows is a smallest 2-coding: $c_0(0) = \Lambda$, $c_0(1) = 0$, $c_0(-1) = 1$, $c_0(2) = 00$, $c_0(-2) = 01$, $c_0(3) = 10$, etc. A corresponding coding can clearly be constructed for any $r > 2$.

Theorem (7.2.5)   If c_0 is a smallest r-coding, and $L_0 = c_0 \bullet \ell$, then for any $c \epsilon C_Q, L = c \bullet \ell$,

$$L(n) \geqslant L_0(n).$$

<u>Proof</u>: Since c is injective and total (definition (7.2.1)) all the elements of $S(n) = \{c(0), c(1), c(-1), c(2), ..., c(n)\}$

exist and are distinct.   Since c is assumed to be size-capturing, we have $|i| < |n| \Rightarrow L(i) \leqslant L(n)$.   Suppose $n \leqslant 0$. Then $|S(n)| = 2|n| + 1$.   Now the number of possible code words (i.e. images of c) of code-length j is r^j, since $c \varepsilon C_Q$.

Hence

$$|S(n)| \leqslant \sum_{j=0}^{L(n)} r^j$$

$$= \frac{r^{L(n)+1} - 1}{r-1}$$

Assume $L(n) < L_0(n) = \left[\log_r |n|\right] + 1$,

i.e. $L(n) \leqslant \left[\log_r |n|\right]$.

Then $|S(n)| \leqslant \dfrac{r^{\left[\log_r |n|\right]+1} - 1}{r-1}$

$\leqslant \dfrac{r|n| - 1}{r-1}$

$= \dfrac{r}{r-1} |n| - \dfrac{1}{r-1}$

$< 2|n| + 1$ since $r > 2$

$= |S(n)|$.

Hence the assumption leads to a contradiction.   If $n > 0$ then the assumption $L(n) < L_0(n)$ leads to the same contradiction since $L(n) = L(-n)$ and $L_0(n) = L_0(-n)$.   Hence $L(n) > L_0(n)$.

The ordinary decimal coding of the integers reserves the symbols $+,-$ for denoting whether the integer is positive or negative.   Consequently its code-length is given by $L(n) = \left[\log_{10} |n|\right] + 2$, if it is assumed that positive integers are

always preceded by "+" (The usual convention of leaving
positive integers unsigned leads to this coding not being
size-capturing, since $L(n) \neq L(-n)$;   furthermore the usual
identification of O with OO,OOO, etc, means that the usual
notation is not a coding according to our definition
(7.2.1)).   Note that since we are interested only in comparing
models, it does not much matter whether the code we use has
length $\left[\log_{10}|n|\right] + 1$ or $\left[\log_{10}|n|\right] + 2$.

7.3   Coding Determines Feature Selection

Suppose that the models being considered have the structure
shown in fig. 9.   Each model has one table look-up, whose
contents are denoted by $E = \{e_i : i=1,\ldots N\}$.   The "fixed"
part of the model, namely that part which is not the table
look-up, is denoted by m.   m operates on the <u>model</u> input
D_i to give $m(D_i)$, which is summed with e_i to give the <u>model</u>
output C_i.   (cf. sec. 3.3).   Suppose further that $C_i = y_i$, the
ith <u>system</u> output observation , and that e_i, y_i are single
integers (as opposed to vectors of integers).   Let $\ell(m)$
be the length of the λ program implementing m, and let $L(E)$
denote $\sum_{i=1}^{N} L(e_i)$, namely the length of the table look-up
when coded into λ (assuming that λ uses a coding c for
which $L = c \bullet \ell$).   Note that we have assumed that any separators
used in the table look-up are included in m.

The suggested assessment procedure for rival models,
namely choosing the shorter model, leads to the selection
of the model with the shorter table look-up, if the number N

of observations is large enough. Thus a change of the coding used for model assessment tends to change the "weighting" which is given to various features of the model behaviour.

For example, if the coding used is such that $L(n) = n^2$ then the model which gives a better least-squares fit will tend to be selected. If $L(n) = |n|^x$ and x is large then the model which gives a better "minimax" fit will tend to be selected.

Suppose that the table look-up of one of the models has $|e_i| = a$, $i = 1, \ldots, N$, whereas another one has $e_i = 0$ for 99% of the entries, and $|e_i| = 3a$ for 1% of the entries. Then the use of a coding with code-length $L(n) = |n|^5$, for example, would tend to favour the second model, while a coding with $L(n) = |n|$ would tend to favour the first. On the other hand, the use of a coding with $L(n)$ increasing sufficiently slowly may be indifferent to either table look-up. Ideally, the coding which should be used is one which best emphasises the features in which the modeller is interested. However, it is not clear whether a coding exists which is capable of reflecting the features which are important to the modeller, or even whether he can say precisely what they are.

The existence of a smallest r-coding strongly suggests that this coding should be used for model assessment. After all, in searching for a good model, the modeller is trying to find the most concise way of expressing, or computing, the observations. It therefore seems natural that he should use the most concise coding.

The use of the smallest r-coding leads to the most
conservative complexity/approximation trade-off, in the
sense that any other size-capturing coding will more quickly
favour a complex model with small fitting errors.   This
results in a "safe" procedure for model assessment, in the
sense that in certain circumstances the results agree with
the results which would be obtained if any other size -
capturing coding were being used.   This will be shown in the
next section.

7.4   Effect of Change of Coding.

Let λ' be a programming language which is identical with
λ, except that integers are coded via the coding c' rather
than c.   If the entries in a table look-up of a λ-model of
a system are transformed by applying the function $c^{-1} \bullet c'$
to them, then the resulting program will be a λ'- model of
the system.   (c^{-1} is a function, since c is injective, by
definition (7.2.1)).   Furthermore, if the λ-model has table
look-up E, and its size is $\ell(m)+L(E)$ (cf. fig. 9), then the
size of the λ'-model is $\ell(m)+L'(E)$, where $L=c\bullet\ell$ and $L'=c'\bullet\ell$.

Suppose that we have two rival models, with "fixed"
parts m_1 and m_2, and table look-ups E_1, E_2, respectively.
Let e_i^1 , e_i^2 denote the entries of E_1, E_2.   Suppose that the
models are assessed using λ and λ'.

Theorem (7.4.1)   Let c,c' be size-capturing codings and
$L=c\bullet\ell$, $L'=c'\bullet\ell$ be such that, for any i,j,

$$|L'(i)-L'(j)|\geqslant|L(i)-L(j)|.$$

If $\ell(m_1)+L(E_1)>\ell(m_2)+L(E_2)$, and a bijection $h:\{1,2,\ldots,N\}\to$
$\{1,2,\ldots,N\}$ exists, such that $|e_i^2|\leqslant|e_{h(i)}^1|$ $(i=1,\ldots N)$,
then $\ell(m_1)+L'(E_1)>\ell(m_2)+L'(E_2)$.

(The bijection h is used to rearrange E_1, so that to every

entry in $h(E_1)$ there corresponds a smaller entry in E_2).

Proof:
$$|L'(e_{h(i)}^1)-L'(e_i^2)|\geqslant|L(e_{h(i)}^1)-L(e_i^2)|$$

But c,c' are size-capturing and $|e_{h(i)}^1|\geqslant|e_i^2|$.

Therefore $L'(e_{h(i)}^1)-L'(e_i^2)\geqslant L(e_{h(i)}^1)-L(e_i^2)$.

So, summing over i, $L'(E_1)-L'(E_2)\geqslant L(E_1)-L(E_2)$.

But $\ell(m_1)+L(E_1)>\ell(m_2)+L(E_2)\implies$

$\quad L(E_1)-L(E_2)>\ell(m_2)-\ell(m_1)$.

Therefore $\quad L'(E_1)-L'(E_2)>\ell(m_2)-\ell(m_1)$,

hence $\quad\quad\ell(m_1)+L'(E_1)>\ell(m_2)+L'(E_2)$.

This theorem, together with the previous discussion,
shows that under the stated condition, if model 2 is better
than model 1 according to a size-capturing coding c, then it
remains better according to any other size-capturing coding
c', whose code-length everywhere increases at least as quickly
as that of c.

The condition stipulating the existence of h can be
expected to be satisfied in many cases of practical interest.

Example (7.4.2)

$\quad E_1=(8,15)$, $E_2=(9,1)$. $(N=2)$

$$h(1)=2, \quad h(2)=1.$$

$$e_1^2 = 9 < 15 = e_{h(1)}^1 \quad , \quad e_2^2 = 1 < 8 = e_{h(2)}^1.$$

So the theorem can be applied, e.g.:

$$L(n) = \log_{10}|n|+1, \quad L'(n)=|n| \quad \text{gives}$$

$$L'(E_1)-L'(E_2)=13>1=L(E_1)-L(E_2).$$

Example (7.4.3)

$$E_1 = (1,15) , \quad E_2 = (8,9). \quad (N=2)$$

In this case a suitable h does not exist, and in fact we have

$L'(E_1)-L'(E_2)=-1<1=L(E_1)-L(E_2)$, so the proof of the theorem

would break down.

Corollary (7.4.4)

If c_O is a smallest r-coding with code-length L_O,

and a bijection $h:\{1,\ldots,N\}\to\{1,\ldots,N\}$ exists such that

$|e_i^2| \leqslant |e_{h(i)}^1|$, $(i=1,\ldots,N)$,

and $\ell(m_1)+L_O(E_1)>\ell(m_2)+L_O(E_2)$, then for any $c \epsilon C_Q$, $L=c \circ \ell$,

such that

$$L_O(|i|+1)-L_O(i)=1 \implies L(|i|+1)-L(i) \geqslant 1,$$

we have

$$\ell(m_1)+L(E_1)>\ell(m_2)+L(E_2).$$

Proof:

From definitions (7.2.3) and (7.2.4),

$$0 \leqslant L_O(|i|+1)-L_O(i) \leqslant 1 \quad , \quad i \neq 0.$$

Also, from definition (7.2.3),

$$L(|i|+1)-L(i) \geqslant 0.$$

These two facts, together with the condition on L imposed in the statement of the corollary, imply that

$$L(|i|+1)-L(i) \geq L_o(|i|+1)-L_o(i).$$

Consequently, for any i,j,

$$|L(i)-L(j)| \geq |L_o(i)-L_o(j)|.$$

The corollary follows from theorem (7.4.1).

Corollary (7.4.4) states that, if model 2 is better than model 1 when a smallest r-coding is used, then it remains better when any other $c \epsilon C_Q$ is used, providing only that L increases at those points at which L_o increases, and that a suitable h can be found. The condition on L does exclude some codings which may conceivably be of interest, such as those for which $L(n) = \max(k, L_o(n))$.

7.5   Summary

A class of codings has been introduced which endows table look-ups of models with the desirable characteristic that their size increases as the magnitudes of their entries increase. A smallest coding exists in this class, for a given size of alphabet. This smallest coding appears to be a natural one to use for model assessment. An advantage of using it is that inmany cases the result of model assessment will agree with the result that would have been obtained with most other codings. The smallest coding is reasonably closely approximated by the conventional r-ary codings of the integers (cf. Bobrow and Arbib (52)).

8   DISCUSSION   AND   CONCLUSION.

8.1   Model Assessment

The method of assessing rival models which we have
proposed has a number of drawbacks.   Does it nevertheless
have practical value? We believe that it does.

The most obvious drawback is that the assessment depends
on the choice of programming language.   As has been mentioned
before, every method of model assessment relies on some
a priori assumptions holding.   Consequently, our method of
assessment is no worse that others in this respect, if the
choice of language is taken as a specification of a priori
assumptions.   Admittedly, this method of specifying such
assumptions is rather indirect, and the type of assumption
which can be specified is somewhat restricted.   For example,
it is not possible to specify a priori statistical knowledge —
which is one of the forms in which a priori knowledge is
commonly specified.   On the other hand, choosing a programming
language is the way in which many a priori assumptions are
actually specified in simulation studies.   In fact, in
discussions of modelling it has become common to associate
choice of simulation language with the modeller's "world view"
(e.g. Mihram (53)).   Furthermore, choices of programming
language can be made which correspond to tighter or looser
specifications of a priori knowledge.

A second drawback is that, although a natural way of

coding table look-up elements has been demonstrated, such a coding exists for any positive integer r⩾2. Which r should be used? Fortunately, there is a smallest possible r (namely 2), and the logarithmic nature of the smallest code-length implies that the relative sizes of models will not change much if r ranges over the commonly used values (say r⩽16). However, the change may be significant; therefore the only meaningful procedure appears to be to try two or three low values of r, including r=2, and see how much this influences the model assessment.

As an example, let us investigate the assessment of the Algol gas-furnace models using binary-coded table look-ups rather than decimal ones. We shall consider only models I, IV and VI. If the elements of the table look-ups of these three models, shown in Appendix C, are each multiplied by 10 to make them integral, and the conventional binary coding applied to the resulting integers (preceded by "+" or "-"), then the assessment shown in table V is obtained. We have assumed that the fixed parts of the models are unchanged, and that the compiler translates the table look-ups back to standard Algol representation.

Model	Size, excluding table look-up	Size of table look-up	Total size	Information gain	Information explained
I	52	3184	3236	0	0
IV	129	1153	1282	1954	60.4%
VI	225	1059	1284	1952	60.3%

Table V - Assessment of Algol Gas-Furnace Models I,IV and VI, Using Binary-Coded Table Look-Ups.

The coding used is not quite a smallest 2-coding,
since we have $L(n) = \left[\log_2 |n|\right] + 2$, but this is of no con-
sequence, since we are interested in <u>relative</u> model sizes.

From table V it is seen that, when a smallest 2-coding
is used , there is virtually no difference between the
performances of models IV and VI. In this case we have
a rather firm indication that model IV is superior to model
VI. Had the ordering of model perforance changed, several
alternative conclusions could have been reached. Firstly,
one could have concluded that models IV and VI are equally
good, in the sense that there is no more support for one
than for the other. Alternatively, if one were trying to
answer the question "Is it worth using a more elaborate
model than model IV?" one could answer "no", since there is
no firm support for model VI. Finally, if one were interested
primarily in a clear-cut decision procedure then one could
choose as better the model which had a higher information
gain when a smallest 2-coding was used.

The two chief advantages of the proposed model assessment
procedures are (1) that it can be applied to complex nonlinear dynamical models, and (2) that the models being
compared do not need to be structurally similar. It is
only necessary to have a means of obtaining the model
behaviour - as was remarked in Chapter 1, this is no great
restriction because of the possibility of simulation.

However, a qualification is necessary here. We have
assumed in this thesis that the models of interest are

essentially deterministic causal processes, with stochastic effects entering in such a manner that they can be considered to be manifestations of the modeller's imperfect understanding of the system.   The models which appear in control theory are of this type.   But in operational research it is common practice to use models which are essentially stochastic in nature, such as the conventional models of queues.   The characterisation of Chapter 3, and consequently the method of assessment based on it, is not immediately applicable to such models.   (In some cases it may be possible to extend the characterisation to such models by requiring the models to compute not the observed behaviour itself, but some observed property of that behaviour - for example, a histogram).

Perhaps the best known type of complex model which can be assessed by our method is the "System Dynamics" type of model introduced by Forrester (6).   Forrester does not in fact subject his models to tests against real data.   He argues that complex industrial and socio-economic systems contain such strong cross-couplings and feedbacks that statistical validation is virtually impossible.   Consequently he insists that the modeller should be able to justify all the detail of the model, and that elements of the model should correspond to "real elements" in the system.   In other words, only those elements are permitted about which the modeller's a priori knowledge is so strong that he has no doubts about their inclusion.   Furthermore, if the model behaves in an unsatisfactory manner, then "causes for the discrepancy must be found which can be explained and defended on grounds other than that their inclusion corrects system behaviour" since "a sufficiently elaborate formal curve-fitting procedure can be devised to fit arbitrarily closely any ensemble of data" (6).

If adhered to strictly, these injunctions are very restrictive. If the modeller is unsure about which of two competing structures should appear in some part of the model, then he should include neither. If, on the other hand, he disregards the above guidelines, and includes one or other structure, then Forrester does not provide him with any means of deciding whether it should in fact be included, or of choosing between the two structures.

The concept of information gain provides a "System Dynamics" modeller with an objective method of making such decisions. To make use of it, the modeller requires some relevant observed data, of course. He can take a model which includes only those structures about which he has no doubt, and assess its information gain. It does not matter what this is - it can even be negative; at this stage the modeller will not have based his model on the particular observations available to him, but on his a priori knowledge - that is, ultimately, on many other observations of similar systems. He can now incorporate one of the competing uncertain structures, and see whether the information gain is improved or not. Obviously, he can also establish which of these structures gives the greater information gain.

Note that this procedure ensures consistency with
Forrester's proviso, that adding new features to a model
should not merely be a "curve-fitting" exercise.   An
increase of information gain indicates that more has been
explained about the system behaviour than has been added to
the model.

The desirability of validating a model by the use of a
set of observations other than the one used for constructing
it is now generally recognised.   Indeed, this can be said
more strongly: the validation of a model by using the same
set of observations as the one used for constructing it is
held to be meaningless.   This is correct if the criterion
used for the validation is goodness-of-fit, since, as has
been stated several times already, an arbitrary goodness-of-fit
can be achieved for any finite set of data.   Now the assessment of information gain is of course a procedure for model
validation, yet only one set of observations is assumed.
Why, then, is it not meaningless?

The reason is that information gain does not just assess
goodness-of-fit, but trades it off against complexity.
The simpler the model, the fewer observations are required
to construct it.   Thus information gain can be regarded
as measuring, in a sense, the "proportion" of the observation
set used for constructing the model, and comparing this with
the "proportion" which successfully validates the model.
An information gain of zero indicates, roughly, that all the
observations have been used to build the model, and none

remain for validation.

Suppose that a certain set of observations is used to construct a model, and that this model is then validated on a second set of observations, in the sense that its behaviour is a good fit to those observations. Then it is clear that the larger the second set of observations is, the greater will be the information gain of the model (viewed as a model of the system which is a concatenation of the two sets of observations). It may be conjectured that the converse is also true: that if a model has a high information gain, then it is possible to divide the observation set into two disjoint sets, such that the model can be constructed using one of the sets, and successfully validated using the other.

Of course, the modeller will inevitably feel safer if he succeeds in validating his model on a second set of data, even if his model does have a high information gain when modelling the first set alone. But if this is not possible (for example, because he would have to wait too long to acquire new data) then the above arguments show that information gain gives an alternative means of model validation.

To summarise, then, it is believed that in spite of certain features which at first appear arbitrary, the proposed method of model assessment can, if used intelligently and with care, be a useful guide to the modeller, insofar as it can indicate to him the success of his efforts to date. The features which remain arbitrary are no more prominent

than the arbitrary features of any model assessment procedure.
The availability of such a guide significantly extends the
range of techniques available to the modeller, for an important
and large class of dynamical models.

8.2   Model Building

The characterisation of modelling which was presented
in Chapters 3 and 4 is of course an idealisation and a
simplification of the modelling process.   Nevertheless, it
exhibits some important characteristics of modelling.   It
differs from other attempted formalisations of modelling
(e.g. (53), (54)) in that it questions some long-cherished
beliefs: namely, that a "true model" necessarily exists of the
system to be modelled, and that there is some inherent virtue
in model complexity.   These beliefs are a legacy from the
modelling of engineering systems, and do not seem appropriate
for socio-economic and other poorly-understood systems.

Most discussions of modelling assume at the outset that
some "true model" of the system under investigation exists,
and that the aim of the modelling exercise is to find a
model which in some sense approximates the true one.   We
make no such assumption.   Consider an asymptotic system
$\mathcal{S} = (S^1, S^2, \ldots)$ (cf. Chapter 4).   Let $x(n)$ be the Gödel number
of the best model of the system S^n (judged according to the
criterion of Chapter 3).   Then it is possible that the
sequence $(x(1), x(2), \ldots)$ is random.   In this case it does
not seem meaningful to speak of the "true model" of $\mathcal{S}$.
However, an asymptotic model of $\mathcal{S}$ can always be found, since

the sequence of trivial models is an asymptotic model.

When modelling engineering situations, in which the detailed cause/effect relationships within systems are well understood, it is reasonable to seek to obtain better models by making them more and more detailed.   Thus an association between complex, detailed models and good models is naturally formed in these circumstances.   In our characterisation precisely the opposite view is taken: other things being equal, good models are small models.   Furthermore, if one cannot build a model using a priori knowledge alone, but must compare its performance with observations, then the size of a credible and useful model is in effect limited by the size of the observation set.   Of course, a priori knowledge is itself ultimately derived from observations, even though the modeller may not be able to exhibit them directly - in a sense, it is a model of such observations. Thus it can be stated quite generally, that the size of useful models is limited by the total amount of observations which have been made, without contradicting the fact that very complex models prove useful under certain conditions, even though they may not be directly supported by large observation sets.   Other treatments of modelling usually pay lip-service to these ideas; however, these notions are not very prominent, and one often finds the author slipping back into a "bigger is better" frame of mind.

For example, Mesarovic and Pestel (55) repeatedly refer to the large size of their model as a point in its favour.   At one point they make the following statement: "In our model about 100,000 relationships are stored in the computer, as compared to a few hundred in other well-known world models".   From the viewpoint of our characterisation, this statement causes grave doubts about the merits of Mesarovic and Pestel's model.   It seems unlikely that enough data could be collected about the "world system" to justify such a large model; it seems even more unlikely that the performance of the model should be so good as to justify the rejection

of a reasonably adequate model which is a thousand times
smaller.   (Although the comparisons would have to be
made with care, since the observation sets would not be
the same for the two models).

Perhaps the most important, albeit not contentious,
result of our characterisation is that in general no
effective procedure for finding the best model of a system
can exist.   This implies that the mechanical application
of some standard system identification technique can not be
guaranteed to produce a useful model, even though it may
be guaranteed to produce an "answer".   This should dispose
of the naive view that "systems analysis" can replace the
specific expertise of the discipline concerned with the
system being studied.   The bounds on what systems
modelling can achieve in a field such as economics depend
primarily on the progress of economic theory, and not on
the technological trappings of modelling and simulation.

8.3   Scientific Inference.

Modelling is concerned with the inference of theories
and hypotheses from observations.   It is therefore interesting
to compare briefly the characterisation of modelling developed
in this thesis with some philosophical studies of general
scientific inference.

Both philosophers of science and scientists have long
regarded simplicity as a desirable characteristic of scientific
theories.   One of the best known expressions of this is the
principle of Ockham's Razor, which states that only those

entities should be introduced into a theory which are absolutely necessary. Newton's "Rules of Reasoning in Philosophy" (56) stem from this principle:

"Rule I: We are to admit no more causes of natural things than such as are both true and sufficient to explain their appearances.

Rule II: Therefore to the same natural effects we must, as far as possible, assign the same causes.

Rule III: The qualities of bodies, which admit neither intensification nor remission of degrees, and which are found to belong to all bodies within the reach of our experiments, are to be esteemed the universal qualities of all bodies whatsoever.

Rule IV: In experimental philosophy we are to look upon propositions inferred by general induction from phenomena as accurately or very nearly true, notwithstanding any contrary hypotheses that may be imagined, till such time as other phenomena occur, by which they may either be made more accurate, or liable to exceptions".

In these rules some similarities with our characterisation can be discerned: rules I and II are an informal counterpart of our emphasis on high information gain, while rules III and IV justify the use of information gain as an indicator of the degree of confidence which may be held in a model.

A detailed attack on the idea that simplicity is desirable has been made by Bunge (57). However, our characterisation

of modelling escapes the brunt of this attack. Bunge's
main concern is that regarding simplicity as the sole criterion
of scientific quality is too naive. Proper emphasis should
also be placed on accuracy and depth: "The motto of science
is not just Pauca but rather Plurima ex paucissimis - the
most out of the least. In short, we wish economy and not
merely parsimony."

Information gain does not just measure the size of a
program. It measures the size of a program, given that the
program is a model of the system. As was mentioned earlier,
this leads to a trade-off between model complexity and
approximation. Thus our characterisation accords with
Bunge's motto. It should be added, however that when modelling
poorly understood systems, the plurima is limited by the nature
of the available observation set; if this is small then
useful models will necessarily have to be small also.

Bunge's second major criticism of a general exhortation
of simplicity is that several different aspects of simplicity
can be discerned, and that these are often mutually incompatible.
An indiscriminating call for simplicity is therefore nonsensical.
We easily evade this criticism, since we have been quite
specific about how simplicity should be measured, and therefore
about what type of simplicity we regard as important.

Our theory concerning models is in some respects a
"realisation" of Popper's abstract theory of scientific method
(38). The first point of agreement is the hypothetico-
deductive nature of the theory. Theorem (3.4.4) shows that

the hypotheses from which our models are built cannot be
obtained in some routine manner - no "universal modelling
algorithm" can exist.   How they are obtained is irrelevant
to us.   Forcing the model to compute the observations
corresponds to the deductive part of the method.   This makes
it possible to test the theory - the size of the corrections
which must be generated by a table look-up gives a measure
of the extent to which the theory is falsified by the
observations.   Alternatively, a table look-up can be viewed
as an ad hoc hypothesis introduced to save the model from being
falsified.   Information gain then measures the empirical content,
or "degree of corroboration", of the overall model:

> "We must not exclude all immunisations, not even all
> which introduced ad hoc auxiliary hypotheses.   ... A
> prima facie falsification may be evaded ... by the
> introduction of testable auxiliary hypotheses, so that
> the empirical content of the system - consisting of
> the original theory plus the auxiliary hypothesis -
> is greater than that of the original system" (58).

Popper's contention that a scientific law which is true
can never be verified, whereas an alleged law which is not
true can be falsified, has its counterpart in our theory.
If a model which has been found is the best (smallest)
possible model of the system under investigation, then we
cannot prove it.   On the other hand, if it is not the best
possible model, then we can demonstrate this by exhibiting
a smaller model.

Our assertion that good models are small corresponds
to Popper's belief that good theories are simple.   Popper

associates "simplicity" with"paucity of parameters", whereas
we have a rather more general concept of simplicity - paucity
of terms in the model.   It is because of this correspondence
that we suggest that information gain can be identified with
Popper's "degree of corroboration".

Kuhn (39) has suggested that science does not in fact
use a uniform method, but that it can be divided into two
distinct phases.   In the usual phase - "normal science" -
fundamental assumptions are not questioned, and routine work
of a "problem-solving" character is pursued.   However, from
time to time a "scientific revolution" occurs - sufficiently
many anomalies and shortcomings of the established "Weltanschauung"
accumulate to force a revision of basic assumptions.   It is
tempting to associate a change of the programming language used
for writing models with such a "scientific revolution".   For,
as has been argued in chapter 4, a change of programming
language implies a change in a priori beliefs about the system
being investigated.   Such a change leads to a change in the
ordering of models, when they are ordered in accordance with
their information gains.   The suggestion that a change in the
ordering of models corresponds to a "scientific revolution" has
previously been made by Gaines ((14) and cf. sec. 2.4.2).

Formal developments of logical probability and of
scientific induction by Carnap and his school (36),(59) always
assume a particular "logical basis".   This is taken to be a
formal language, in which all scientific statements could in
principle be made.   In our characterisation of modelling a
corresponding assumption is made, namely the assumption of a
particular programming language.   A major criticism of the
assumption of a formalised language has always been that it
is evident that scientific statements are never expressed in

such languages, and that it is not certain whether a formal
language capable of expressing interesting scientific
statements can exist.   This contrasts with the status of the
programming languages which we have to assume.   Obviously,
such languages are capable of expressing statements which
are interesting to the modeller, and in many cases it is
practical to do so.   Furthermore, it has been shown in
chapter 5 that such languages can be defined quite precisely.

A tenuous similarity can be pointed out between our
characterisation and Carnap's "Continuum of Inductive Methods"
(60).   Carnap introduces a "confirmation function", which
is to be interpreted as the logical probability that a particular
event will occur.   The value of this function depends on a
term which can be interpreted as an a priori logical factor, and
on an a posteriori empirical factor, which is a relative
frequency.   The relative weighting of these two factors is
governed by a positive real parameter λ, which thus indexes
the "continuum of inductive methods".   The value of the
parameter λ is usually taken to be chosen subjectively, and
depends on how regular or "lawlike" the inductivist believes
his "universe of discourse" to be.

Somewhat analogously, we can consider our characterisation
of modelling to be an "enumeration of inductive methods", which
is indexed by a Gödel number of the programming language
which has been chosen.   The choice of this index is also
subjective, since it reflects a priori beliefs.   However,
we are not proposing that our method leads to a logical
probability.   Although it is easy to obtain a $[0,1]$-valued

function from information gain, it does not seem possible
to make it behave like a probability. The essential difficulty
is that no effective way of normalising the function can
exist. (The function would have to be additive, and would
need to sum to unity over a set of models which is not
effectively computable).

Finally, we point out that our characterisation of
modelling escapes the two famous paradoxes of induction
(61),namely those of Hempel and Goodman. We include consideration
of these because they tend to be regarded as "benchmark"
tests of accounts of inductive inference. Hempel's paradox
arises as follows: suppose it is hypothesised that "all
ravens are black". This hypothesis is equivalent to the
hypothesis that "all non-black things are non-ravens".
Any theory of induction which is based on "confirming instances"
supporting a hypothesis seems to lead to the conclusion, that
every observation of a non-black thing which is also not a
raven supports the original hypothesis that "all ravens are
black". This is patently absurd.

Our characterisation clearly does depend on "confirming
instances" in a sense, because every time a hypothesised regularity
is repeated in the system behaviour, the information gain of
the model is increased. However, Hempel's paradox does not
arise, because although a model can be viewed as the statement
of a hypothesis, in general no model will exist which corresponds
to the logical negation of that hypothesis (we can write an
algorithm of the form x:=y; but not x:$\neq$y;). Thus we have
a very different notion of what type of entity can be "supported"
by observations from the classical one -such entities cannot

be merely logical statements, but must be algorithms.

The second paradox is Goodman's "grue" paradox, which
also arises from consideration of "confirming instances".   Every
time a green emerald is observed, the hypothesis that
"emeralds are green" is confirmed, and the hypothesis that
"emeralds are blue" is falsified.   However, the observation
also confirms the hypothesis that "emeralds are grue" - namely,
green until 1980 and blue thereafter, and falsifies the
hypothesis that "emeralds are bleen" - blue until 1980 and green
thereafter.   Thus the observation would appear to tell us nothing
at all about the appearance of emeralds in the future.

This paradox is evaded, rather than solved, by our
characterisation.   We assume that a particular programming
language has been chosen.   The definition of this language
can be regarded as the definition of what basic predicates
are to be used in our scientific statements.   If basic
predicates like "blue" and "green" are defined, then predicates
like "grue" and bleen" can be constructed from these - it is
helpful to think of them as being defined by procedures.
Now a theory like "emeralds are green" can be expressed as
a model by using just the terminal characters of the language.
However, a model corresponding to "emeralds are grue" would
need to include the declaration of the procedure "grue";
consequently, its information gain would be lower, and this
model would be rejected in favour of the first one.

Of course, this makes no contribution to the philosophical
problem of why the language chosen should define "blue" and
"green" rather than "grue" and "bleen".

8.4   Systems Science

It has long been perceived that control theory and,
more generally, systems theory are principally concerned
with the acquisition, transfer and use of information,
rather than of energy.   Models are often said to convey
information.   However, "information" in this sense is
usually used in an intuitive, presystematic way.

Information gain has been introduced in an attempt to
formalise this idea of information conveyed by a model.   The
established theories of information do not appear adequate
for this purpose.   Use of the statistical theory of information
transmission would have restricted us to the consideration
only of random processes which could be described statistically,
and the ideas of "cause" and "effect" could have found no place
in such a framework.   On the other hand, Carnap and Bar-
Hillel's theory of semantic information (59), (62) would have
involved the use of uncomputable "logical probabilities",
and would in any case not be capable of practical application
to modelling.

The algorithmic information theory of Kolmogorov (19)
almost provides what is required, since it defines information
not in terms of probabilities, but in terms of algorithms
- and models are clearly algorithms.   Consequently,
information gain has been developed from the ideas of algorithmic
information theory.   However, it must be emphasised that
information gain is not the same entity as Kolmogorov's
"quantity of information".   The latter is an uncomputable

quantity, whereas it has been deliberately ensured that
information gain is computable. Furthermore, Kolmogorov's
"quantity of information" is a function defined on pairs of
signals, whereas information gain is a function defined on
models (assuming the "signals" to be given).

How can the assertion that information gain measures
the performance of a model be justified? We have employed
four arguments in its favour. The first is argument by
association: algorithmic information is a plausible
formalisation of "information"; information gain is intimately
related to algorithmic information; therefore information
gain is a plausible measure of "information". The second
is argument by weight of opinion: many of the workers who
have attacked similar problems have been drawn to similar
conclusions (e.g. Wrinch and Jeffreys, Solomonoff, etc. cf.
chapter 2). The third argument we have used is the argument
of consistency with what we would expect: there is no
universal method for finding the best model; for large
observation sets models are chosen on the basis of goodness
of fit; for large observation sets a priori beliefs do not
matter; for an important class of processes, the same model
is chosen as best, as would be chosen by established theory.
The fourth argument is that of "operationalism": the information
gain of models which are of practical interest can be calculated.
This has been demonstrated by an example.

A stronger type of argument than any of these would be
the use of the concept of information gain to obtain new
results in systems theory, or to explain known phenomena of

systems modelling.   Rissanen has already provided one
result of this kind ((41) and cf. sec. 2.5).   In effect,
Rissanen has shown that if the information gain of the model
of a Gauss-Markov process is maximised, then the resulting
identification scheme is an extension of standard maximum
- liklihood techniques.

A second possible application of information gain is to
the explanation of what Young has termed the "Law of Large
Systems" (63).   This is the conjecture, based on experience,
that complex, poorly understood systems can often be adequately
represented by very simple models.   One can see immediately
how the explanation of this in terms of information gain
would run.   The possible information gain of a model of such
a system is limited by the size of the available information
sets.   Suppose that models (understood informally) belonging
to a certain class are fitted to these observation sets in
order of increasing complexity.   Let t be the size of the
trivial model, and I_n be the information gain of the nth
model so fitted when suitably formalised.   Then in passing
from the nth to the (n+1)th model, the greatest possible
improvement in information gain is $t-I_n$.   If t is small, it
seems likely that diminishing returns would quickly set in,
and no improvement in information gain would be possible
after the first few models had been fitted.   It would be
interesting to establish the conditions under which this
does indeed happen.

8.5   Conclusion

A characterisation of modelling has been introduced
which is believed to incorporate certain salient features of
the modelling of complex, poorly understood systems, such
as those which are encountered in environmental, socio-economic
and management studies, and in certain industrial control studies.
This characterisation consists of three parts: a system, a
model, and a criterion of model quality.

A system is considered to be defined by a pair of
observation sets - an input set and an output set.   Each
observation set is assumed to be a finite array of rational
numbers.

A model is an algorithm which computes the output
observation set.   It may use certain elements of the observation
sets to help it in this task, as specified by the modeller.
The model is not allowed to approximate the system behaviour,
but must compute it exactly.   The trivial model is a model
which represents the situation at the beginning of the modelling
exercise, before any structure has been discerned in the
system behaviour.

The criterion of model quality is the model's information
gain.   This is the amount by which the model is smaller
than the trivial model, when both are written as a computer
program.   In conventional terms, this criterion leads to a
trade-off between model complexity and model accuracy.   This
prevents "overfitting" of the model to the observations, and
puts a premium on finding the greatest amount of regularity

in the system behaviour.

The ranking of models according to this criterion depends on the programming language chosen for writing the models. However, this is not entirely arbitrary, because it can be associated with the modeller's *a priori* beliefs about the system. The manner of coding the observations also affects the model ranking. Again, this is not arbitrary, because a distinguished smallest coding has been shown to exist, and this coding is a natural one to use.

A detailed investigation has shown that the notion of "*a priori* information assumed by a model" can be precisely defined, by associating it with the smallest language required to write the model as a program. However, an examination of the conditions under which models can be meaningfully compared has indicated that it is of no great consequence if such conditions are not met exactly.

The work reported in this thesis serves two purposes. Firstly, it provides a formal analysis of modelling which is logically sound, and which gives a plausible formalisation of the concept of "information supplied by a model". Secondly, it gives the modeller a simple technique for assessing the progress of his modelling efforts. Our guiding idea has been that, other things being equal, complexity in a model indicates vacuousness rather than sophistication.

<u>REFERENCES</u>

1. Weber, R.L, "A Random Walk in Science", The Institute of
    Physics, (1973), p.92.

2. Astrom, K.J. and Eykhoff, P, "System Identification - A
    Survey", Automatica, 7, (1971), 123-162.

3. Akaike, H, "Autoregressive Model Fitting for Control",
    Annals of the Institute of Statistical Maths, 23, (1971),
    163-180.

4. Chan, C-W, Harris, C.J, and Wellstead, P, "An Order-Testing
    Criterion for Mixed Autoregressive Moving Average Processes",
    Int. J. Control, 20, (1974), 817-834.

5. Akaike, H, "A New Look at the Statistical Model Identification",
    IEEE Trans. Auto. Control, AC-19, (1974), 716-723

6. Forrester, J.W, "Industrial Dynamics", M.I.T. Press and Wiley,
    (1961).

7. Von Neumann, J, and Morgenstern, O, "The Theory of Games and
    Economic Behaviour", Princeton, (1944).

8. Fuller, A.T., "Analysis of Nonlinear Stochastic Systems by
    Means of the Fokker-Planck Equation", Int. J. Control, 9,
    (1969), 603-655.

9. Rogers, H, "Theory of Recursive Functions and Effective
    Computability", McGraw-Hill, (1967).

10. Box, G.E.P. and Jenkins, G.M, "Time Series Analysis,
    Forecasting and Control", Holden-Day, (1970).

11. Kalman, R.E, Falb, P.L, and Arbib, M.A, "Topics in Mathematical
    Systems Theory", McGraw-Hill, (1969).

12. Windeknecht, T.G, "General Dynamical Processes", Academic
    Press, (1971).

13. Zadeh, L.A, and Desoer, C.A, "Linear System Theory-The
    State Space Approach", McGraw-Hill, (1963).

14. Gaines, B.R, "System Identification, Approximation and
    Complexity", Int. J. Gen. Systems, 3, (1977), 145-174.

15. Blum, M. "On the Size of Machines", Info & Contr, 11, (1967),
    257-265.

16. Blum, M. "A Machine-Independent Theory of the Complexity of
    Recursive Functions", J. ACM, 14, (1967), 322-336.

17.   Löfgren, L, "Complexity of Descriptions of Systems",
      Research Report IVl 7601, Dept. of Automata and General
      Systems Sciences, Lund Institute of Technology, (January 1976).

18.   Hartmanis, J, and Hopcroft, J.E, "An Overview of the Theory
      of Computational Complexity", J. ACM, 18, (1971), 444-475.

19.   Kolmogorov, A.N, "Three Approaches to the Quantitative
      Definition of Information", Problems of Information Transmission,
      1, No. 1, (1965), 1-7.

20.   Kolmogorov, A.N,"Logical Basis for Information Theory and
      Probability Theory", IEEE Trans. Info. Theory, IT-14,
      (1968), 662-664.

21.   Zvonkin, A.K., and Levin, L.A., "The Complexity of Finite
      Objects and the Development of the Concepts of Information
      and Randomness by Means of the Theory of Algorithms",
      Russian Mathematical Surveys, 25, no. 6, (1970), 83-124.

22.   Solmonoff, R.J., "A Formal Theory of Inductive Inference",
      Information and Control, 7, (1964), 1-22, and 224-254.

23.   Chaitin, G.J., "On the Length of Programs for Comptuing
      Finite Binary Sequences," J.ACM, 13, (1966), 547-569.

24.   Martin-Löf, P., "The Definition of Random Sequences",
      Information and Control, 9, (1966), 602-619.

25.   Church, A, "On the Concept of a Random Sequence", Bull
      Amer. Math. Soc., 46, (1940), 130-135.

26.   Gillies, D.A., "An Objective Theory of Probability",
      Methuen,(1973)

27.   Schnorr, C.P."Optimal Enumerations and Optimal Gödel
      Numberings", Math. Syst. Th, 8, (1975), 182-191.

28.   Meyer, A.R, "Program Size in Restricted Programming Languages",
      Info& Contr., 21, (1972), 382-394.

29.   Biermann, A, & Feldman, J.A, "A Survey of Grammatical
      Inference" in: Watanabe, S, (ed), "Frontiers of Pattern
      Recognition", Academic Press, (1972), 31-54.

30.   Fu, K.S, & Booth, T.L, "Grammatical Inference: Introduction
      and Survey-Part I", IEEE Trans Syst., Man & Cybernetics,
      SMC-5, (1975), 95-111.

31.   Gold, M, "Language Identification in the Limit", Info &
      Contr., 10, (1967), 447-474.

32.   Chomsky, N, "On Certain Formal Properties of Grammars",
      Info & Contr., 2, (1959), 137-167.

33. Fu, K.S, & Booth, T.L, "Grammatical Inference: Introduction and Survey-Part II", IEEE Trans. Syst, Man & Cybernetics, SMC-5, (1975), 409-423.

34. Feldman, J, "Some Decidability Results on Grammatical Inference and Complexity", Info & Contr., 20, (1972), 244-262.

35. Blum, L, and Blum, M, "Toward a Mathematical Theory of Inductive Inference", Info & Contr., 28, (1975), 125-155.

36. Carnap, R, "Logical Foundations of Probability", University of Chicago Press, (1950).

37. Wrinch, D, and Jeffreys, H, "On Certain Fundamental Principles of Scientific Inquiry", Philosophical Magazine, ser. 6, vol. 42, (1921), 369-390

38. Popper, K.R., "The Logic of Scientific Discovery", Hutchinson, (1959).

39. Kuhn, T.S, "The Structure of Scientific Revolutions", University of Chicago Press, (1962).

40. Löfgren, L, "Relative Explanations of Systems" in Klir, G.J, "Trends in General Systems Theory", Wiley, (1972), 340-407.

41. Rissanen, J, "Parameter Estimation by Shortest Description of Data", Proceedings of JACC Conference, ASME, (1976), 593-597.

42. Rissanen, J, "Basis of Invariants and Canonical Forms for Linear Dynamic Systems", Automatica, 10, (1974), 175-182.

43. Chaitin, G.J. "A Theory of Program Size Formally Identical to Information Theory", J.ACM, 22, (1975), 329-340.

44. Eykhoff, P, "System Identification-Parameter and State Estimation", Wiley, (1974).

45. Box, G.E.P., and Jenkins, G.M, "Time Series Analysis, Forecasting and Control", Holden-Day, (1970).

46. Whittle, P, "Prediction and Regulation by Linear Least-Squares Methods", EUP, (1963).

47. Sherman, S, "Non-Mean-Square Error Criteria", IRE Trans. Info. Theory, IT-4, (1958), 125-126.

48. Johnston, J, "Econometric Methods", McGraw-Hill, (1963).

49. Ollongren, A, "Definition of Programming Languages by Interpreting Automata", Academic Press, (1974).

50.   Challis, M.F, "Algol W Language Specification and Programmer's
      Guide", University of Cambridge Computing Service, 3rd
      edition (August 1975).

51.   Deutsch, R, "Estimation Theory", Prentice-Hall, (1965).

52.   Bobrow, L.S, and Arbib, M.A, "Discrete Mathematics:
      Applied Algebra for Computer and Information Science",
      W.B. Saunders, (1974).

53.   Mihram, G.A, "Simulation: Statistical Foundations and
      Methodology", Academic Press, (1971).

54.   Zeigler, B.P, "Theory of Modelling and Simulation",
      Wiley, (1976).

55.   Mesarovic, M, and Pestel, E, "Mankind at the Turning Point",
      Hutchinson, (1975).

56.   Cajori, F, "Sir Isaac Newton's Mathematical Principles of
      Natural Philosophy and his System of the World, vol. 2",
      University of California Press (1962).

57.   Bunge, M, "The Myth of Simplicity", Prentice-Hall, (1963)

58.   Popper, K.R, "Unended Quest", Fontana/Collins, (1976).

59.   Hintikka, J, and Suppes, P, "Information and Inference",
      Reidel, (1970).

60.   Carnap, R, "The Continuum of Inductive Methods", University
      of Chicago Press, (1952).

61.   Hesse, M, "The Structure of Scientific Inference", Macmillan
      (1974).

62.   Bar-Hillel, Y, "Language and Information", Addison-Wesley
      and The Jerusalem Academic Press, (1964).

63.   Young, P.C, Shellswell, S.H, and Neethling, C.G., "A
      Recursive Approach to Time Series Analysis", Cambridge
      University Engineering Department, Technical Report CUED/B-
      Control/TR16(1971).

64.   De Bakker, J.W. "Semantics of Programming languages" in:
      Tou, J.T. (ed), "Advances in Information Systems Science,
      vol. 2", Plenum Press, (1969), pp.173-227.

65.   Lauer, P. "Formal Definition of Algol 60", IBM Laboratory,
      Vienna, Technical Report TR 25.088, (1968) .

66.   Zimmermann, K. "Outline of a Formal Definition of Fortran",
      IBM Laboratory, Vienna, Technical Report LR.25.3.053, (1969).

67.   Neuhold, J. "The Formal Description of Programming Languages",
       IBM Systems Journal, 10, 2, (1971), pp.86-112.

68.   Lucas, P., Lauer, P, Stigleitner, H, "Method and Notation
       for the Formal Definition of Programming Languages" IBM
       Laboratory, Vienna, Technical Report TR 25.087, (1968),
       (revised 1970).

69.   Aho, A.V. & Ullman, J.D. "The Theory of Languages", Mathematical
       Systems Theory, 2, 2, pp 97-125, (1969).

APPENDIX A

Formal Semantics of Programming Languages.

A.1   Introduction

The formal definition of the semantics of a programming language is a notoriously difficult problem of computer science (64).   Fortunately, a lot of progress has been made in recent years, so that several solutions now exist.   In this appendix one of these solutions, the so-called Vienna Method, will be described.   This method of defining semantics has been chosen because it is the best documented, and because it has been used for the definition of several practical programming languages (65), (66).   There need therefore be no doubts about its power.

A very complete and careful account of the Vienna Method is given by Ollongren (49), but more concise, and in some ways clearer, descriptions are available in (67) and (68). Only a brief introduction can be given here.   This intro-duction will be illustrated by formally defining a special language, called "Linear Model Language", or LML.   This language is chosen for exposition because it is very simple - too simple, in fact, to need most of the sophistication of the Vienna Method.

The Vienna Method involves the definition of a language in four stages.   First, the concrete syntax of the language is defined.   This specifies those strings of characters which are valid programs in the language.   The specification

is invariably in Backus-Naur form; the concrete syntax is
therefore a context-free grammar (69). The concrete syntax
indicates how each string of characters which forms a program
is to be parsed. On completion of parsing certain characters
(such as semicolons and comment strings in Algol) become
redundant, and can be discarded. The remaining entities
which appear on the nodes of the parsing tree are now mapped
into entities to which semantic roles will eventually be
assigned. This mapping is specified by defining a translator;
this is a function which maps the structured object which is
a (parsed) concrete program into another structured object
called an abstract program. The set of structured objects
which are valid abstract programs of the language is called
the abstract syntax of the language. The language definition
is completed by defining an interpreting automaton. This is
defined by specifying a set of structured states of the
automaton, and a state transition function. It is the
definition of this function which really determines how
programs in the language are to be interpreted. A computation
is viewed as a sequence of states of this automaton. The
initial state is determined by the program and its data.

Tree-structured objects are all-pervasive in the Vienna
Method. They are used to represent both the concrete and
abstract syntax of a language, and to represent the states
of the interpreting automaton. A special function, called
the μ-function, is introduced to carry out "tree-surgery",
and is used extensively.

Finally, it is assumed that a metalanguage is available which can be used for the definition of the above entities.

A.2   Linear Model Language

Suppose that we wish to investigate only single-input, single-output systems, whose input and output values are rational numbers, and whose input-output relation is a finite-order difference equation, together with an additive random disturbance of the the output. Let u_i be the ith input observation of such a system, y_i its ith output observation, and d_i the ith random disturbance. Let a_j, b_j be coefficients of the difference equation, so that the following equation holds:

$$y_i = a_1 y_{i-1} + \ldots + a_n y_{i-n} + b_o u_i + \ldots b_m u_{i-m} + d_i.$$

We can (informally) define a programming language as follows. Every program of the language is a list of integers and rationals which is given the interpretation: $n, m, a_1 a_2, \ldots, a_n b_o, b_1 \ldots, b_m, d_1, d_2, \ldots, d_N$. The data for such a program is a similar list, with the interpretation $i, y_{i-1}, \ldots, y_{i-n}, u_i, \ldots, u_{i-m}$. Given such a program and such a set of data, the computation of y_i in accordance with the above equation is invoked. If input observations $(u_{-m}, \ldots, u_N)$, and output observations $(y_{-n}, \ldots, y_N)$ of a system are obtained, then a certain (infinite) set of programs in this language will constitute models of the system defined by the observations $((u_1, \ldots, u_N), (y_1, \ldots, y_N))$. The terms $d_1, \ldots, d_N$ which appear

in the program are, in fact, a look-up table.   Fig. 7
shows the structure of each such model.

The programming language described above will be
called Linear Model Language, or LML.   Note that LML is
not a universal language, in the sense that not every
algorithm can be implemented in it.

A trivial model of a system is obtained in LML if
$m=n=b_o=0$.

A.3   Tree-Structured Objects and the μ-Function.

As mentioned earlier, tree-structured objects are much
used in the Vienna Method.   We assume here that such objects
are familiar.   The following example should clarify their
nature, and will also introduce some terminology and notation.

A typical tree-structured object (or simply "object",
for short), is the following entity A:

A   =

s_1, s_2, s_3, are called <u>simple selectors</u>.   The objects appearing
at each node of the tree are themselves tree-structured,
with the objects e_1, e_2, e_3, which appear at the terminal

nodes of the tree, being regarded as degenerate trees.
These objects are called <u>elementary objects</u>.   The object B
is some tree-structured object which is not necessarily

elementary.

Every object has a finite number of nodes, and each node, other than the unique <u>root</u>, is associated with a unique <u>preceding node</u>.   A successor node n_j of a preceding node n_i is"selected" by a simple selector s, and this is denoted by $n_j = s(n_i)$.   Strictly speaking, n_i and n_j here denote objects rather than nodes.   Thus, in the example, we have $e_1 = s_1(A)$. The simple selectors associated with any node must be pairwise distinct.   This makes it possible to select the object associated with any node by composing simple selectors. For example, in the above object we have $e_2 = s_1 \bullet s_2(A)$, $e_3 = s_2 \bullet s_2(A)$, $B = s_3 \bullet s_2(A)$, where $\bullet$ denotes composition of selectors.   Entities of the form $s_i \bullet \ldots \bullet s_j$ are called <u>composite selectors</u>.   Note that reading a composite selector from left to right corresponds to "reading" an object from bottom to top.

The <u>null object</u> is associated with the empty tree, that is the tree with no nodes, and is denoted Ω.

Let K denote a composite selector, and e an elementary object.   Then the <u>characteristic set</u> of an object A is the set of all pairs $\langle K:e \rangle$, such that $K(A) = e$.   An object can be defined by giving its characteristic set.   For example, the above object A is defined by:

$$A = \{\langle s_1:e_1 \rangle, \langle s_1 \bullet s_2:e_2 \rangle, \langle s_2 \bullet s_2:e_3 \rangle, \ldots\} \ .$$

The characteristic set of B is not known in this case, so this definition cannot be completed.   But suppose that B were the object:

$B=\{<s_1:e_4>,<s_2:e_5>\}\ =$

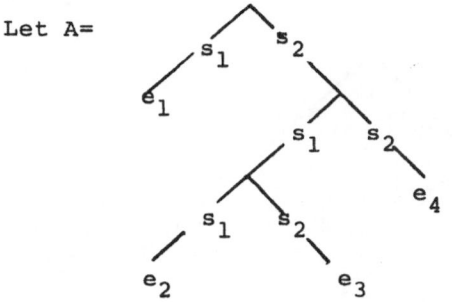

then we would have

$A=\{<s_1:e_1>,<s_1\bullet s_2:e_2>,<s_2\bullet s_2:e_3>,<s_1\bullet s_3\bullet s_2:e_4>,<s_2\bullet s_3\bullet s_2:e_5>\}.$

We now introduce the μ-function, which is used to perform operations on objects. The μ-function takes two arguments, the first of which is an object A, and the second is a pair<K:B>, where K is a composite selector, and B is an object. The range of μ is the set of all objects. The value μ(A;<K:B>) is an object which is obtained from A by replacing K(A) by B in such a way that K(μ(A;<K:B>))=B. This is most clearly shown by examples (taken from Lucas et al (68)):

Let A=

Then

(i)   μ(A;<s_3:B>)=

(ii) $\mu(A;<s_1\circ s_2:B>) =$

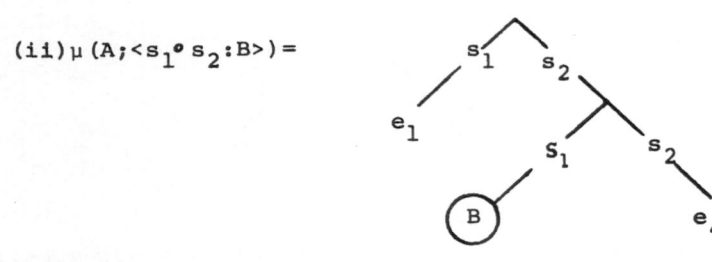

(iii) $\mu(A;<s_1\circ s_1\bullet s_1\bullet s_2:B>) =$

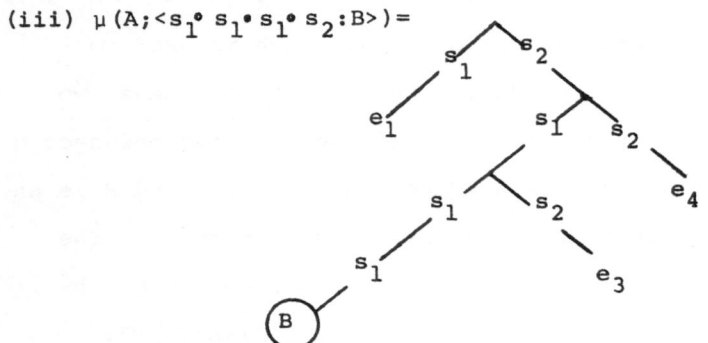

In particular, if $B=\Omega$, we obtain:

(i) $\mu(A;<s_3:\Omega>) = A$

(ii) $\mu(A;<s_1\bullet s_2:\Omega>) =$

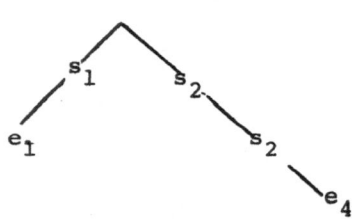

(iii) $\mu(A;<s_1\bullet s_1\circ s_1\circ s_2:\Omega>) =$

Ollongren (49) and Lucas et al (68) define the μ-function more precisely than has been done here.

We now introduce the following notations:

(i) $\mu(A;<K_1:B_1>,<K_2:B_2>,\ldots,<K_n:B_n>) \triangleq$

$\qquad \mu(\mu(A;<K_1:B_1>);<K_2:B_2>,\ldots,<K_n:B_n>)$,

with $\mu(A;) \triangleq A$.

Example

Let A=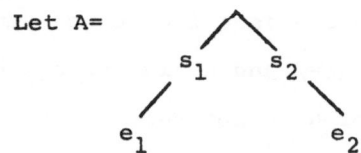

Then $\mu(A;<s_3:e_3>,<s_1 \bullet s_2:e_4>) =$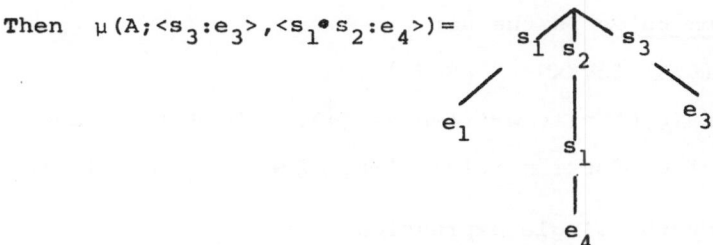

Ollongren (49) gives conditions under which interchanging the arguments of the μ-function leaves the value unchanged.

(ii) $\mu_o(<K_1:B_1>,\ldots,<K_n:B_n>) \triangleq \mu(\Omega;<K_1:B_1>,\ldots,<K_n:B_n>)$

Thus μ_o is a function which "creates" objects.

Example

$\mu_o(<s_1:e_1>,<s_1 \bullet s_2:B>,<s_2 \bullet s_2:e_3>) =$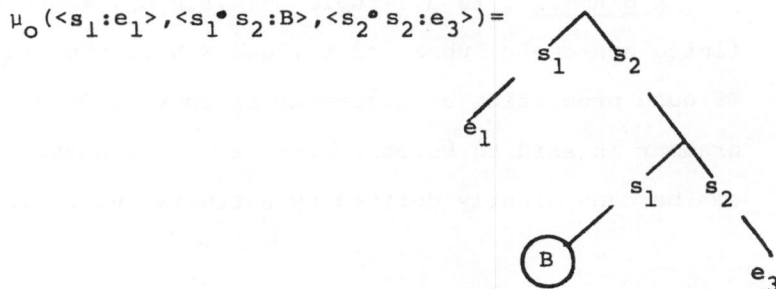

A.4   Concrete Syntax.

The concrete syntax of a programming language can be defined by using the Backus-Naur form of writing production rules.   This is a shorthand method of defining a grammar. Suppose that there exists a finite non-empty set Σ of terminals.   Typical elements of Σ are: b, 2, *, begin, and so on.   Let Σ^* denote the set of all finite strings of elements of Σ.   Also, suppose N is a finite non-empty set of non-terminals such that $N_\cap \Sigma=\emptyset$ and N^* is the set of all finite strings of elements of N.   Let $V=\Sigma \cup N$, and $V^*=(\Sigma \cup N)^*$. Let $V^+=V^*-\Lambda$, where Λ is the empty string.   Then the set of production rules is the set

$$\pi=\{(\xi,\beta):\xi \epsilon V^* x N x V^* \ \& \beta \epsilon V^+\}.$$

Each pair $(\xi,\beta) \epsilon \pi$ is written as $\xi \to \beta$.   In Backus-Naur form, the set of production rules $\{\xi \to \beta_1 , \xi \to \beta_2 ,..., \xi \to \beta_n\}$ is denoted by the single expression

$$<\xi>::=\beta_1|\beta_2|..|\beta_n.$$

The brackets <> are used to denote non-terminals.   Backus-Naur notation can be used to express production rules (ξ,β), providing that $\xi \epsilon N$.   Such production rules are called context-free.

A grammar G is a 4-tuple $G=(N,\Sigma,P,S)$, where P is a finite non-empty subset of π , and $S \epsilon N$ is the start symbol. If each production of a grammar is context-free then the grammar is said to be context-free.   A context-free grammar can be conveniently defined by a finite set of expressions in Backus-Naur form.

If there exist $\delta_1, \delta_2 \epsilon$ V* and $\xi \rightarrow \beta \epsilon P$ such that $\gamma_1 = \delta_1 \xi \delta_2$ and $\gamma_2 = \delta_1 \beta \delta_2$ then $\gamma_1 \underset{G}{\Rightarrow} \gamma_2$. If $\gamma_1 \epsilon$V* and $\gamma_{i-1} \underset{G}{\Rightarrow} \gamma_i$ (i=1,2,...,n), then $\gamma_0 \underset{G}{\overset{*}{\Rightarrow}} \gamma_n$ (γ_n is derived from γ_0).

The grammar G is said to generate the language
$$L(G) = \{x : S \underset{G}{\overset{*}{\Rightarrow}} x \ \& \ x \epsilon \Sigma^* \}$$

Two grammars are equivalent if they generate the same language.

The Vienna Method defines two grammars for each language. One is the concrete syntax, the other the abstract syntax. Lucas et al (68) explain the distinction most clearly: "An abstract syntax is one which only specifies the expressions of the language as to the structures significant for their subsequent interpretation and not as to how they are to be expressed for the purpose of communication either to oneself or to others. A concrete syntax specifies the expressions of the language as a set of character strings".

The concrete syntax of LML is defined as follows:
```
<program> ::= <integer>, <integer>, <rational>, <rational>, ...,
                                                 <rational>.
<rational>::= +<number>| -<number>|O
<number>  ::= <integer>| <integer> .<integer>|.<integer>
<integer> ::= <digit>|< integer><digit>
<digit>   ::= 0|1|2|3|4|5|6|7|8|9
```
The terminals of LML are : 0 1 2 3 4 5 6 7 8 9 . , + -

Positive rationals are required to be signed so that the

terms in the table look-up will be coded in a **size-capturing**
manner (cf. chapter7).   This requirement is extended to
the other terms solely for simplicity of definition.
An example of a valid string in LML is: 2,1,+.6,-3,0,-1.41,-5.2.
This program can be parsed(using the syntax definition) to
give the object (not shown in full):

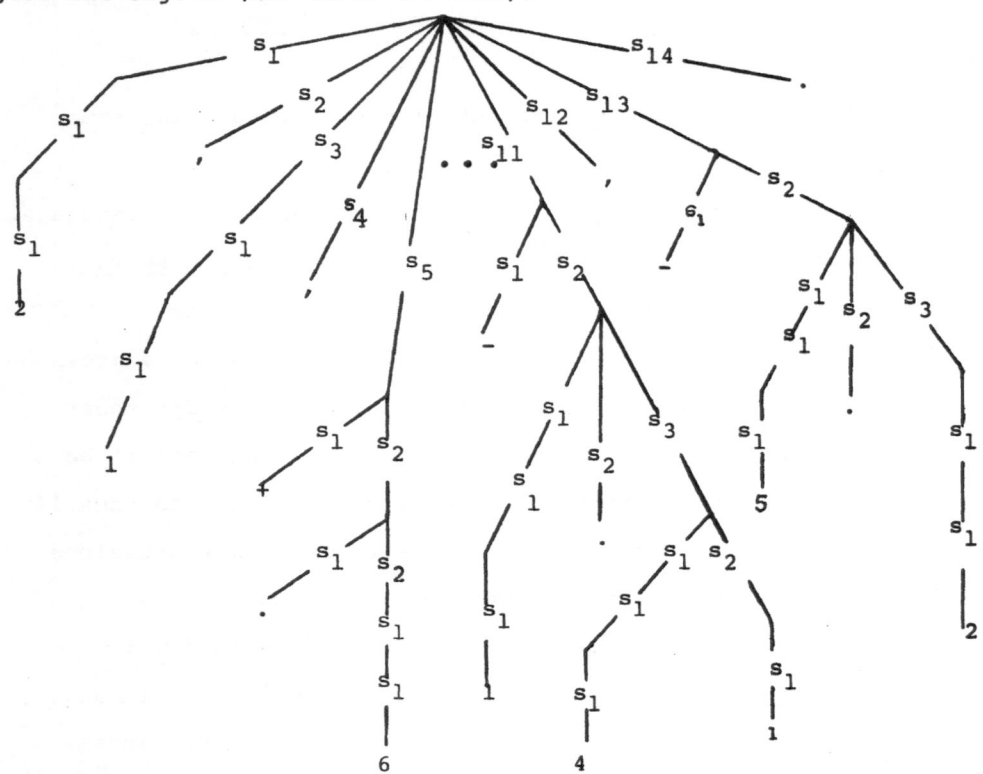

A.5.   Abstract Syntax

We introduce the following notational conventions:
if an object x satisfies a predicate P, we write is -P(x).
The set of objects which satisfy P is denoted is-$\hat{}$P.
That is, is-$\hat{}$P={x:is-P(x)} .   The set is-$\hat{}$P is defined by

an expression of the form

$$\text{is-}P= (<\text{s-}P_1:\text{is-}P_1>,<\text{s-}P_2:\text{is-}P_2>,\ldots<\text{s-}P_n:\text{is-}P_n>)$$

which indicates that for every $x \in \text{is-}\hat{P}$,

$$x=\mu_o(<\text{s-}P_1:x_1>,<\text{s-}P_2:x_2>,\ldots,<\text{s-}P_n:x_n>),$$

where $x_1 \in \text{is-}\hat{P}_1$, $x_2 \in \text{is-}\hat{P}_2,\ldots,x_n \in \text{is-}\hat{P}_n$. If is $-P=$

$(<\text{s-}P_1:\text{is-}P_1>)$ then we write $\text{is-}P=\text{is-}P_1$. A predicate may

also be defined by using the disjunction operator V, e.g.:

$\text{is-}P=\text{is-}P_1$ V $\text{is-}P_2$, which denotes that $x \in \text{is-}\hat{P}$ only if

$x \in \text{is-}\hat{P}_1 \cup \text{is-}\hat{P}_2$.   It is assumed that certain predicates

are satisfied by subsets of the elementary objects.

Using this notation, the abstract syntax of LML is

defined as follows:

is-program=(<s-n: is-integer>,<s-m: is-integer>,<s-rational-

list: is-rational-list>)

is-rational-list=(<s-head: is-rational>,<s-tail:-is-rational-

list V is $-\Omega$>)

It is assumed that $\text{is-}\hat{\Omega}=\{\Omega\}$, and that the predicates is-

integer and is-rational are satisfied by (countably)

infinite sets of elementary objects.   Every LML "abstract

program" satisfies the predicate is-program.   For example,

the abstract program corresponding to the concrete LML

program introduced in section A.4 is the object

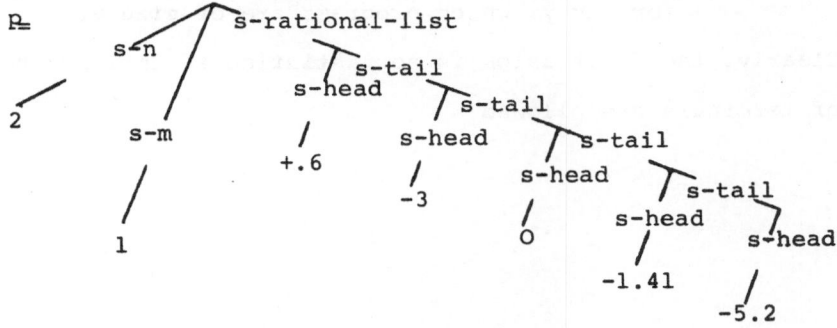

How this object is obtained from the concrete program is
specified by the translator, which will be defined in the
next section.   Note that writing "+.6" for the object
s-head • s-rational-list (p) does not attach any meaning to
that object.".6" is merely an arbitrary label for that
object, although it is chosen to have mnemonic value.

Most discussions of the Vienna Method point out that
the abstract syntax of a language can be viewed as its
defining syntax, and that alternative concrete realisations
of it need not be considered to be separate languages.
If we were to adopt this view, then the above abstract
syntax would not be satisfactory for our purposes, since it
assumes an infinite set of "terminals" (and is therefore
not a "grammar" as defined in section A.4).   It is essential
for us to have a measure of the size of programs, and the
measure introduced in chapter 3 is the length of the string
of terminals which constitutes a program.   As discussed
earlier Blum (15) has introduced a useful pair of axioms
which must be satisfied by a size measure over "machines"
(in our case, over programs).   These are:

(1)   there exists at most a finite number of programs
      of any given size,

(2)   there exists an effective procedure for deciding,
      for any y, which programs are of size y.

Clearly, the first axiom is not satisfied if infinite sets
of terminals are allowed.

Furthermore, we want programs to describe effective procedures for computing functions. Defining a language with infinitely many terminals would correspond to defining a Turing machine with infinitely many tape symbols. This would be a fundamantal change in the notion of "computability".

To overcome these objections, it would be possible to define an abstract syntax for LML which specified a finite set of terminals. The object at each terminal node of an abstract program would then satisfy one of the predicates is-digit or is-sign, say, and these would be defined by

is-digit= is-0 V is-1 V...V is-9

is-sign= is-+ V is- -,

and $\hat{is-}0=\{0\}$, ..., $\hat{is-}9=\{9\}$, $\hat{is-}+=\{+\}$, $\hat{is-}-=\{-\}$.

In this case the assembly of the digits into integers and rationals would have to be performed by the interpreting automaton, rather than by the translator.

A.6   The Translator

The translator is a function which maps the set of parsed concrete programs in a language into the set of abstract programs. To distinguish between concrete and abstract objects we introduce the conventions: is-<program>(x) means that x is a parsed concrete program, namely an object such as that shown in section A.4. More precisely, for LML we have, for some positive integer k:

is-<program>=(<s_1:is-<integer>>,<s_2:is-,> ,...<s_{2k-1}:is-<rational>>,

<s_{2k}:is-.>)

The predicates is-<integer>, is-<rational> are obtained from the concrete syntax in exactly the same way.   Obviously, we have is-$\hat{}$,={,}, is-$\hat{0}$={O}, etc.

In the following definition, the statement __if__...__then__ ...__else__... is a statement in the metalanguage.   It is assumed that is-<program>(p) and is-<rational>(x_1).   The LML translator, trans-program, is defined as: trans-program (p)=

μ_0(<s-n:trans-integer (s_1(p))>,

<s-m:trans-integer (s_3(p))>,<s-rational-list: makelist(s_5(p),

$$s_7(p),\ldots,s_{2k-1}(p))>)$$

where

makelist ($x_1,x_2,\ldots,x_n$)=μ_0(<s-head:trans-rational(x_1)>,

<s-tail:__if__ $x_2=\Omega$&...&$x_n=\Omega$ __then__ Ω __else__ makelist ($x_2,\ldots,x_n$)>)

and the functions

trans-rational: is-$\hat{}$<rational> → is-$\hat{}$rational

trans-integer : is-$\hat{}$<integer> → is-$\hat{}$integer

are not further defined.   For our purposes these two functions are best thought of as the usual mappings onto the rational numbers.   (In an actual implementation, it may be more useful to consider them as mappings into bit-patterns.   In this case the sets is-$\hat{}$rational and is-$\hat{}$integer would be finite sets, due to the fixed word-length of practical computers).

Note that trans-program (p) ε is-$\hat{}$program, and makelist ($x_1,\ldots,x_n$) ε is-$\hat{}$rational-list.

A.7   The Interpreting Automaton

Following Ollongren (49), we define an interpreting
automaton to be a 5-tuple (O, is-state, ξ_o, Λ, F), where
O is the set of tree-structured <u>objects</u> already introduced,
and is-state is a predicate over O.   Objects satisfying
is-state are <u>states</u> of the automaton. ξ_o ε is-$\hat{\text{state}}$
is the <u>initial state</u> of the automaton, and F is a set of
final states.   Λ is the <u>state transition function</u>;
however, its range is not is-$\hat{\text{state}}$, but the power set of is-$\hat{\text{state}}$.
Λ(ξ) is thus a <u>set</u> of states in general, although in our
definition of LML, Λ(ξ) will always be a single state.

A.7.1 The State

The state of the interpreting automaton is structured.
The structure depends on the language to be defined, and for
the definition of LML can be rather simple.   A language with
block structure, procedures, variable identifiers of various
types, conditional and <u>goto</u> statements, and so on, will
need a rather more complicated set of states.

For the LML interpreting automaton, or <u>LML machine</u>, we
define
is-state= (<s-c: is-c>,<s-dn:is-dn>,<s-counter:is-integer>).
is-dn is a predicate satisfied by a <u>denotation directory</u>,
and is defined by
is-dn=(<s-data:is-data>,<s-y: is-rational>,<s-parno: is-integer

V is -Ω>)

where

is-data= (<s-i:is-integer>,<s-list:is-rational-list>).

The data for a program, namely the sequence $i,y_{i-1},y_{i-2},\ldots$

appears in the initial state as the object

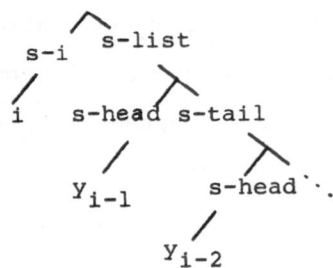

We do not specify how this is achieved. Similarly, the
result of the computation, y_i, is the object s-y•s-dn(ξ_F),

where $\xi_F \in F$, and we do not specify how it is output. The
number m+n, which is required for the correct interpretation
of the program, is stored in s-parno • s-dn (ξ). (The term
"denotation directory" is taken over from (49) and (68).
For LML this directory is simpler than in (49) and (68),
but it serves essentially the same purpose, namely storing
intermediate and final results).

The most complex part of the state is the <u>control</u>,
which is an object satisfying the predicate
is-c= (<s-in: is-in>,<s-al: is-obj-list>,

        <s-ri: is-dum V is-Ω>,<r:is-c>) V is-Ω,
where the following abbreviations have been used:
c: control, in: instruction, al: argument list, obj : object,
ri: return information, dum: dummy.

In this definition, is-in is a subset of the elementary

objects, called the set of <u>instructions</u>, and is-dum is a
subset of the elementary objects called the set of <u>dummy names</u>.
r is a simple selector, different from s-in, s-al or s-ri.
is-obj-list is a predicate satisfied by lists of objects,
which we do not define further; Ollongren (49) gives an
extensive discussion of lists.

An example of a control is the object:

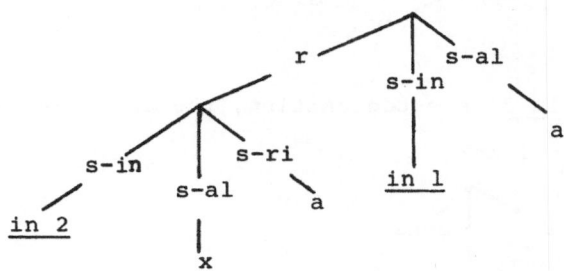

This particular control may have the following effect:
the instruction <u>in 2</u> is performed, with x as its argument.
The result of carrying out <u>in 2</u> is assigned to the dummy
name a. in·2 is then deleted, so that the control part
of the next state is

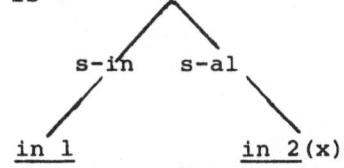

<u>in 1</u> is now carried out, with <u>in 2</u>(x) as its argument.
In this case, <u>in 2</u> is said to be <u>contracting.</u> On the
other hand, it may be that carrying out <u>in 2</u> requires first
carrying out some other instruction <u>in 4</u> on x, and then an
instruction <u>in 3</u> on <u>in 4</u> (x). In this case <u>in 2</u> is said to

be <u>expanding</u>, and carrying it out results in the next state having the control:

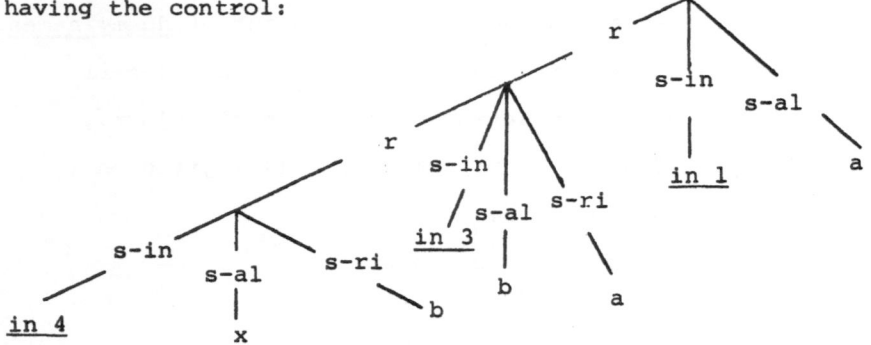

If both <u>in 4</u> and <u>in 3</u> are contracting, the controls of the next two states will be:

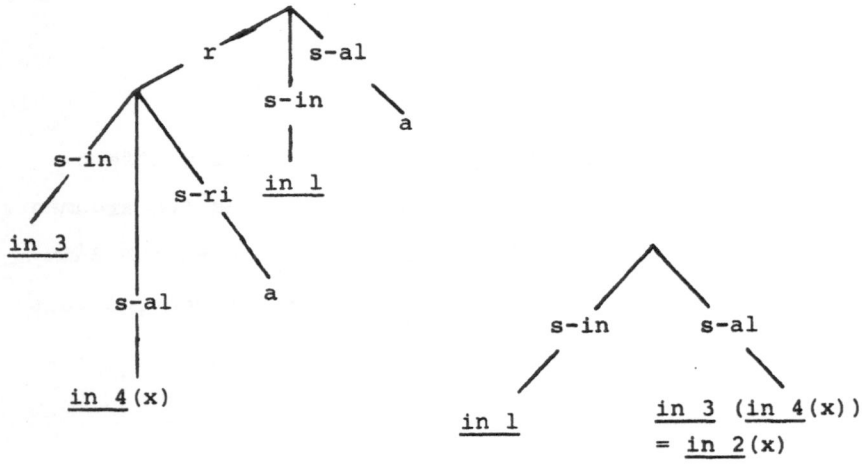

If an instruction is expanding, then performing it leaves all components of the state unchanged except the control itself. However, if it is contracting, then its effect may be to change any of the components of the state (in our case, s-counter (ξ) and s-dn (ξ), as well as s-c(ξ)).

We need some definitions for later use. The set of

<u>control selectors</u> of a control C is the set of composite selectors $\kappa(C) = \{K : K = r \circ r \circ \ldots \circ r \ \& K(C) \neq \Omega\}$, if $C \neq \Omega$, and is I (the identity selector) if $C = \Omega$. The <u>terminal control selector</u> of a control C is the composite selector

$$T(C) = \{\tau : \tau \epsilon \kappa(C) \ \& r \circ \tau \notin \kappa(C)\}.$$

If $K = r^n$ is a control selector of a non-empty control, where $r^n = r \circ r \circ \ldots \circ r$ (n compositions), and $n \geqslant 1$, then the <u>mth</u> <u>preceding control selector</u> of K is

$$\text{prec}^{\ m}(K) = \{K' : r^m \circ K' = K\} \quad (0 \leqslant m \leqslant n).$$

If K is a control selector of a non-empty control C, then

$$\text{prec-arg } (C,K) = \{\text{s-al} \circ \text{prec}^1(K) : i \geqslant 1 \ \& \ \text{s-al} \circ \text{prec}^1(K)(C)$$
$$= \text{s-ri} \ \circ \ K(C) \neq \Omega\}$$

is the set of composite selectors which select those arguments of instructions associated with preceding control selectors of K which are equal to the dummy name associated with K.

If K is a control selector of a non-empty control C then the <u>derived return information</u> associated with K is the set $\text{ri}(C,K) = \text{prec-arg } (C,K)$. (This definition is included because these two sets differ in (49), but coincide for the relatively simple LML machine).

The <u>initial state</u> of the LML machine is

$$\xi_0 = \mu_0(\text{<s-c:}\mu_0(\text{<s-in: }\underline{\text{int-prog}}\text{>,<s-al:p>})\text{>,<s-counter: 1>,}$$
$$\text{<s-dn:}\mu_0 (\text{<s-data: d>,<s-y: 0>})\text{>)}.$$

Here, p is the LML program, which satisfies the predicate is-program introduced in section A.5. The object d

satisfies the predicate is-data defined earlier in this
section.   int-prog is an instruction which will be defined
later.

The set of final states of the LML machine is the set
$F=\{\xi:is\text{-}state(\xi)\ \&\ s\text{-}c(\xi)=\Omega\}$.

A sequence $\xi_0,\xi_1,\ldots$, where $\xi_{i+1}\varepsilon\Lambda(\xi_i)$ is a computation of
the LML machine.   If, for some i, $\xi_i\varepsilon$ F then the computation
terminates.   (Every LML computation terminates).

A.7.2   The State Transition Function

With every instruction in ε is-$\hat{\text{in}}$ is associated an
interpreting function Φ_{in}.   Let C be a non-empty control
of a state ξ, and K a control selector of C.   Let s-in•K(C)=in,
and let ARG= s-al•K(C) be the list of arguments of in.
Then

$$\Phi_{in}(ARG,\xi,K) = \underline{if}\ P_1(ARG,\xi)\ \underline{then}\ \ g_1$$

$$\underline{else}\ \underline{if}\ P_2(ARG,\xi)\ \underline{then}\ \ g_2$$

$$\underline{else}\ .\ .\ .$$

$$\underline{else}\ \underline{if}\ P_m(ARG,\xi)\ \underline{then}\ \ g_m,$$

where $P_1,P_2,\ldots,P_m$ are predicates (m$\geq$1), and g_j has one of
two forms:

(i)   For the case of contracting control,

$$g_j=\mu(\mu(\mu(\xi;<K\bullet s\text{-}c:\Omega>);\{<\tau\bullet s\text{-}c:\varepsilon_o^j(ARG)>:\tau\varepsilon ri(C,K)\});$$

$$<s\text{-}counter:\ is\text{-}integer>,<s\text{-}dn:\varepsilon_1^j(ARG)>)$$

where ε_o^j andε_1^j are objects.   In this expression the innermost μ
deletes the instruction in,its argument list and its return

information, the middle μ <u>returns</u> the object $\varepsilon_{\circ}^{j}$(ARG) to preceding control selectors, and the outermost μ alters components of the state other than the control.

(ii)   For the case of expanding control,

$$g_{j} = \mu(\xi; <K \bullet s\text{-}c: \mu(\varepsilon^{j}(ARG); <s\text{-}ri: s\text{-}ri \bullet K \bullet s\text{-}c(\xi)>)>),$$

where ε^{j}(ARG) satisfies the predicate is-c.   In this case the inner μ associates the return information of K(C) with the new control ε^{j}(ARG), and the outer μ replaces the control K(C) with the new object thus created.

The Vienna Method uses a system of <u>instruction schemata</u> to define interpreting functions rather more concisely and in a more readable fashion than the above expressions. However, we shall not describe this feature, since it is feasible to define LML in the above manner.

It is now possible to define the <u>state transition function</u>:

$$\Lambda(\xi) = \{\eta: \eta = \mathbf{\underline{z}}_{in}(ARG, \xi, K) \ \& \ K = T(s\text{-}c(\xi)) \ \& \ \underline{in} = s\text{-}in \bullet K \bullet s\text{-}c(\xi)$$

$$\& \ ARG = s\text{-}al \bullet K \bullet s\text{-}c (\xi)\}.$$

From this definition it is apparent that the state transition is determined by always carrying out the instruction associated with the <u>terminal</u> control of the state, namely the instruction occurring at the "deepest" level of the control (cf. examples in section A.7.1).   In general (although not for LML), $T(s\text{-}c(\xi))$ will be a set containing more than one control selector.   Hence our earlier remark that $\Lambda(\xi)$ will in general be a set of states, rather than a single state.   In such a case, it does not matter which of the terminal instructions is performed first.

It is the specification of the interpreting functions
of an interpreting automaton which assigns meaning to an
abstract program.

A.7.3   Interpreting Functions for LML

We now complete the definition of LML by defining a
set of interpreting functions for it.   The instructions to
be defined are as follows:

Instruction	Type	Domain
int-prog	expanding	is-$\overset{\wedge}{\text{program}}$
int-mn	expanding	$(\text{is-}\overset{\wedge}{\text{integer}})^2$
set-mn	contracting	is-$\overset{\wedge}{\text{integer}}$
int-prog-list	expanding	is-$\overset{\wedge}{\text{rational}}$-list
updatey	contracting	is-$\overset{\wedge}{\text{rational}}$
product	contracting	$(\text{is-}\overset{\wedge}{\text{rational}})^2$
sum	contracting	$(\text{is-}\overset{\wedge}{\text{rational}})^2$

We assume that the binary arithmetic operators + and * are
available.   The remarks at the end of section A.6 apply
to these.

(1)   int-prog

$$\Phi_{\text{int-prog}} (p, \xi_o, I) = \mu(\xi_o; < s\text{-}c: \varepsilon(p)>)$$

where $\varepsilon(p)$ =

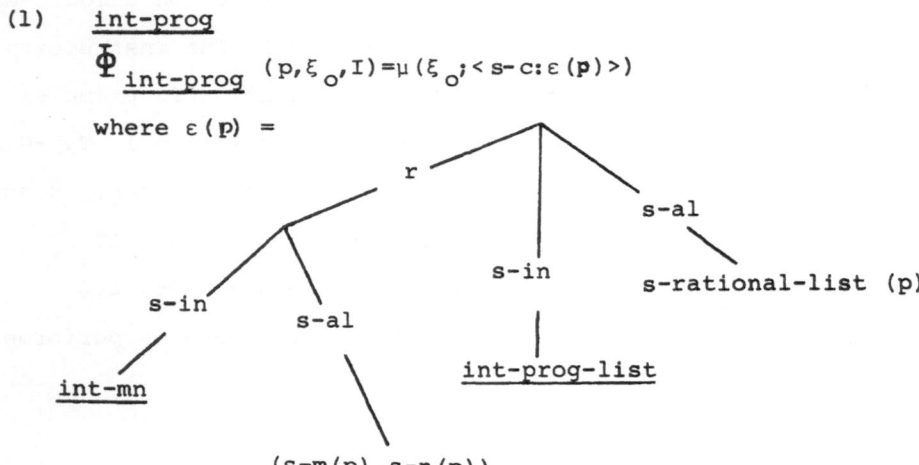

(2)   <u>int-mn</u>

$\underline{\Phi}_{\underline{int-mn}}((x,y),\xi,K)=\mu(\xi;\ <K\bullet s\text{-}c:\epsilon(x,y)>)$

where $\epsilon(x,y) =$

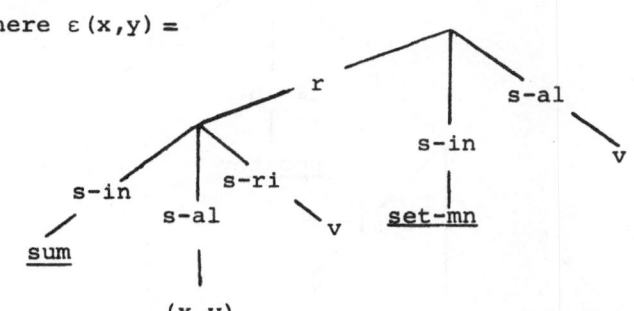

(3)   <u>set-mn</u>

$\underline{\Phi}_{\underline{set-mn}}(x,\xi,K)=\mu(\mu(\xi;<K\bullet s\text{-}c:\Omega>);<s\text{-}dn:\mu(s\text{-}dn(\xi);<s\text{-}parno:x>)>)$

Note: <u>set-mn</u> puts the value m+n into s-parno•s-dn(ξ).

(4)   <u>int-prog-list</u>

$\underline{\Phi}_{\underline{int-prog-list}}(x,\xi,K)=$ <u>if</u> s-counter $(\xi)<$s-parno•s-dn$(\xi)+2$

<u>then</u> $\mu(\xi;<K\bullet s\text{-}c:\epsilon^1(x)>)$

<u>else</u> $\mu(\xi;<K\bullet s\text{-}c:\epsilon^2(x)>)$

where $\epsilon^1(x) =$

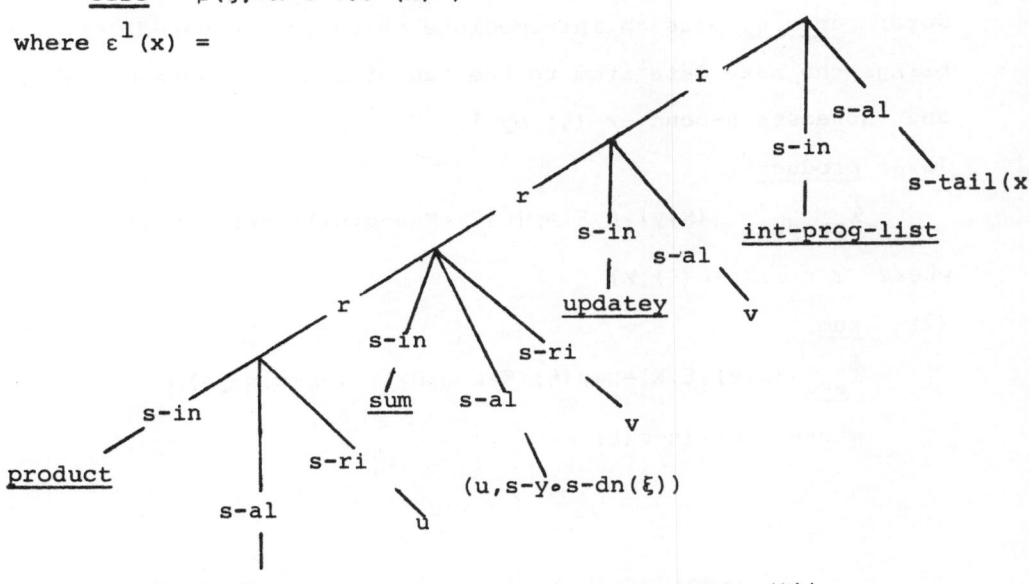

and $\varepsilon^2(x)$ =

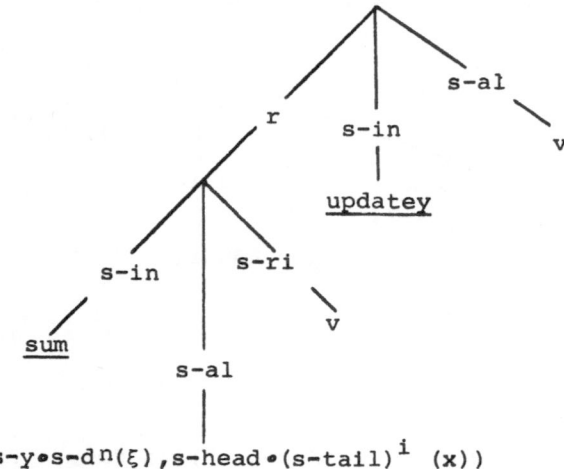

$$(s\text{-}y \bullet s\text{-}d^n(\xi), s\text{-}head \bullet (s\text{-}tail)^i(x))$$

where $i = s\text{-}i \bullet s\text{-}data \bullet s\text{-}d^n(\xi)$

(5) updatey

$$\Phi_{\underline{updatey}}(x,\xi,K) = \mu(\mu(\xi;<K \bullet -c:\Omega>);<s\text{-}d^n:\mu(s\text{-}d^n(\xi);<s\text{-}y:x>,$$

$$<s\text{-}list \bullet s\text{-}data: s\text{-}tail \bullet s\text{-}list \bullet s\text{-}data \bullet s\text{-}d^n(\xi)>),$$

$$<s\text{-}counter: s\text{-}counter(\xi)+1>)$$

Note: updatey puts an intermediate value into $s\text{-}y \bullet s\text{-}d^n(\xi)$,
brings the next data item to the top of $s\text{-}list \bullet s\text{-}data \bullet s\text{-}d^n(\xi)$,
and increases $s\text{-}counter(\xi)$ by 1.

(6) product

$$\Phi_{\underline{product}}((x,y),\xi,K) = \mu(\mu(\xi;<K \bullet s\text{-}c:\Omega>);<\tau \bullet s\text{-}c:x*y>)$$

where $\tau \in ri(s\text{-}c(\xi),K)$

(7) sum

$$\Phi_{\underline{sum}}((x,y),\xi,K) = \mu(\mu(\xi;<K \bullet s\text{-}c:\Omega>);<\tau \bullet s\text{-}c:x+y>)$$

where $\tau \in ri(s\text{-}c(\xi),K)$

In order to clarify the above definitions, some of the steps in an LML computation are shown below. To save space, only the control and those parts of the state which have just changed are shown.

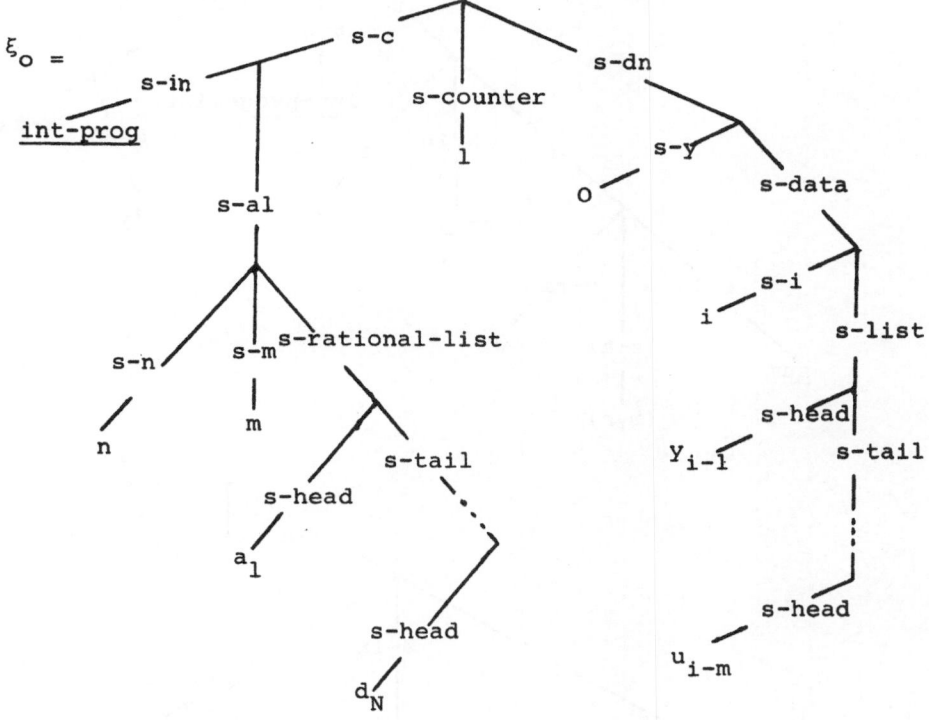

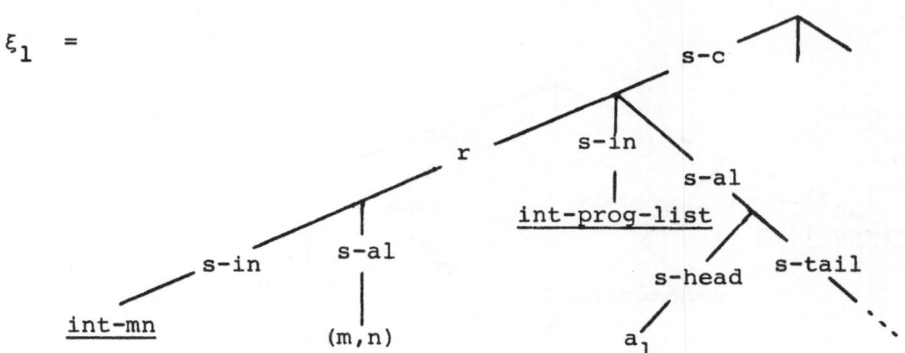

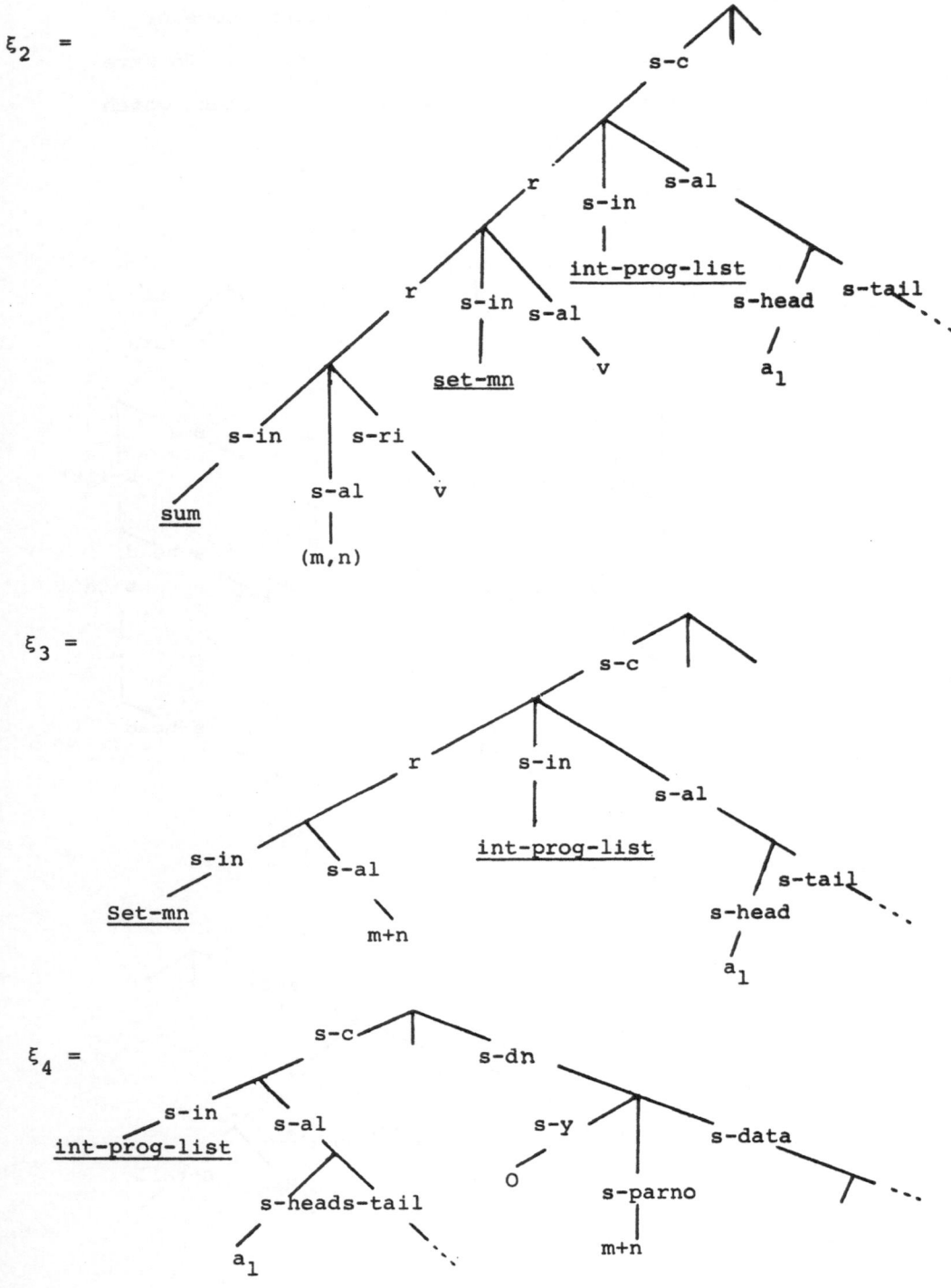

if m+n > 0 then ξ_5 =

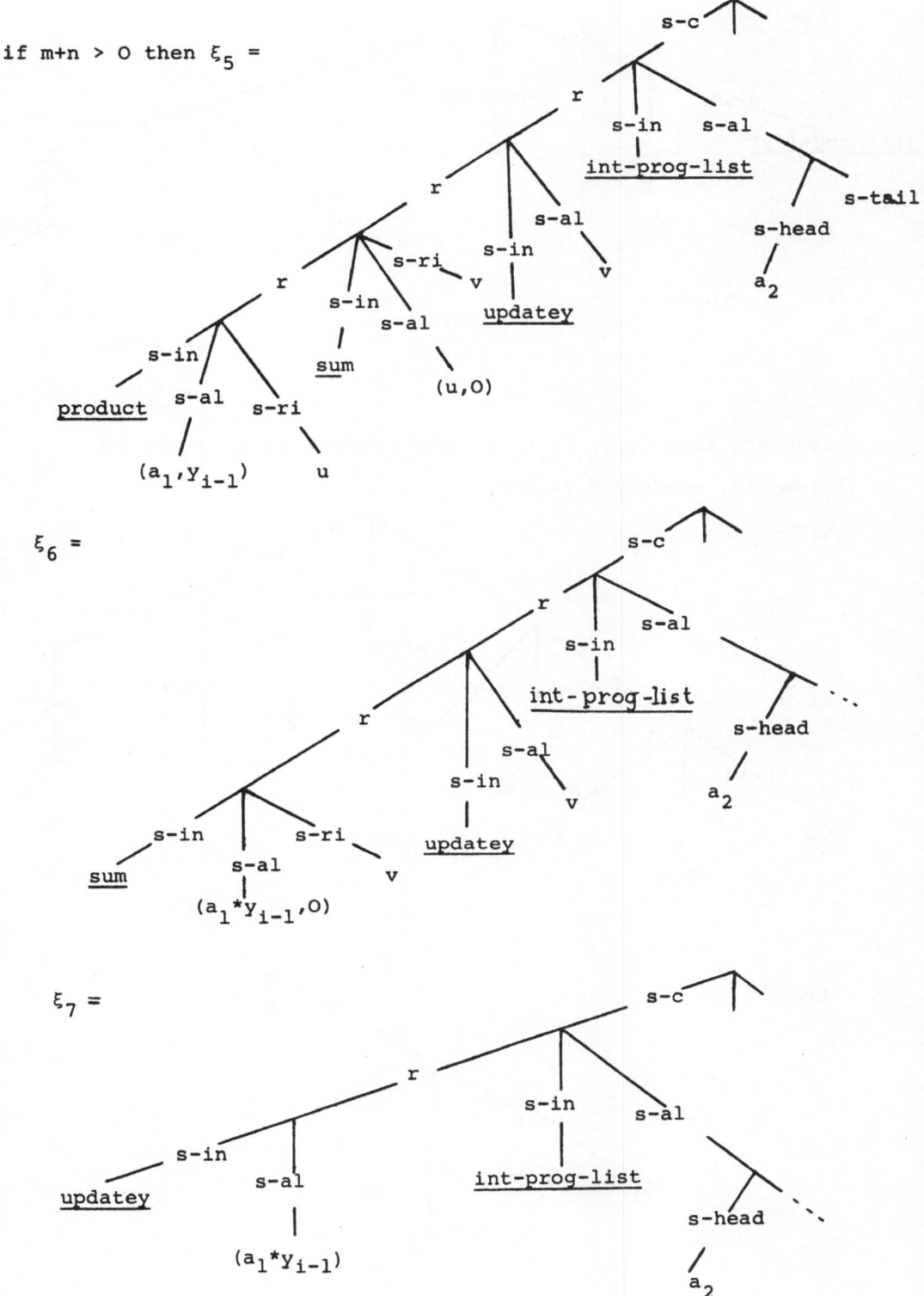

ξ_6 =

ξ_7 =

$\xi_8 =$

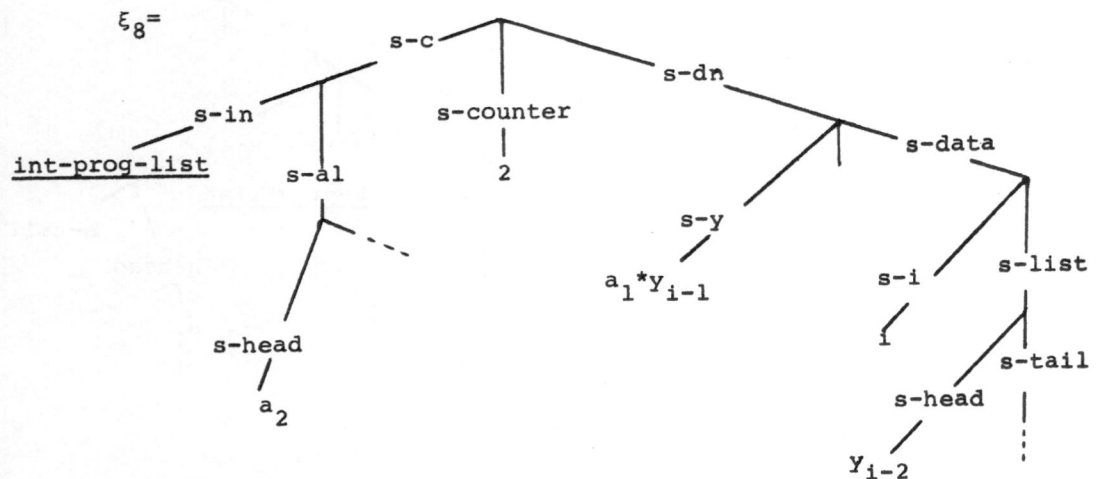

A sequence like $\xi_5, \xi_6, \xi_7, \xi_8$ is now repeated until s-counter $(\xi_i) = m+n+2$, whereupon we get

$\xi_{i+1} =$

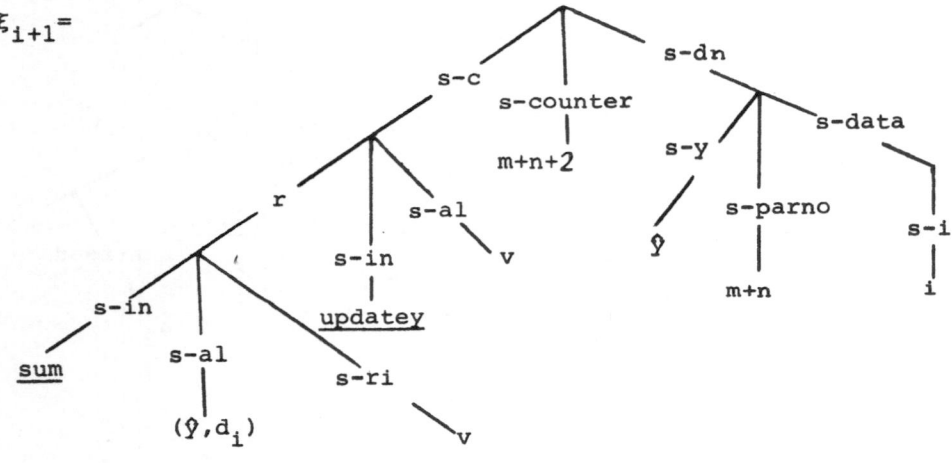

$\xi_{i+2} =$

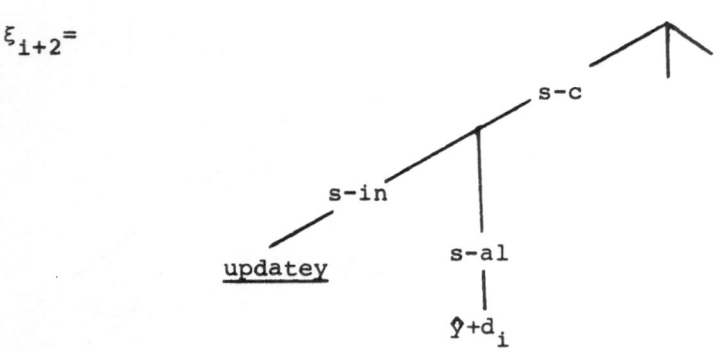

$\xi_{i+3} =$

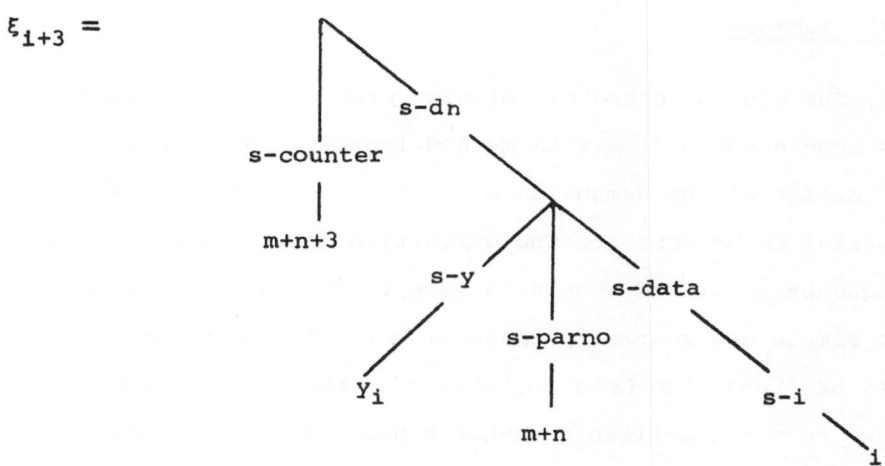

$\xi_{i+3}\epsilon F$, so the computation has terminated.   Its result is available in s-y•s-dn(ξ_{i+3}).

It should be remarked that the above definitions of the LML instructions are not quite complete.   There are restrictions on an LML program which cannot be expressed in the earlier specifications of concrete and abstract LML grammars.   These are, that the number of parameters and the number of data items be compatible with the values of m and n, and that the value of i should not exceed N, the length of the look-up table.   These restrictions can be expressed in the definitions of the LML instructions.   This is simply done by entering an error state if any of these conditions are violated.   In languages like Algol, this technique can also be used to specify context - sensitive restrictions, which cannot be expressed in the context-free grammar definitions (see (49)).

A.8   Summary

The Vienna method of defining programming languages
has been described.   This method includes the formal
definition of the semantics of a language, and is sufficiently
powerful to be used for the definition of practical programming
languages.   It has been used here for the definition of
the simple and special-purpose Linear Model Language.
This has been done both to illustrate the method, and in
order to make familiar a rather broader notion of "programming
language" than is usual.

The Vienna Method of language definition is used in
chapter 5 to formalise the notion of a "fragment" of a language.

APPENDIX B

Syntax of the AlgolW-Support of the Gas-Furnace

Models

This appendix contains the concrete syntax of the
AlgolW-support of the five models of section 6.3.2.   It
is based on the AlgolW syntax specification given in (50).
The numbers in brackets to the right of subheadings indicate
the relevant sections of (50), in order to facilitate
comparison.   Standard procedure statements are new non-
terminals which do not appear in (50) (cf. sec. 6.3.1).
The symbol "t" may be replaced by either "real" or "integer",
in accordance with the rules specified in sections 1.1,
1.5, 1.5.3, and 1.6.2 of (50).

1.   Identifiers (1.2)

```
<identifier>::= <letter>
<t variable identifier>::=<identifier>
<t array identifier>::=<identifier>
<standard procedure identifier>::= READ|READON|WRITE
<letter>::= E|I|J|N|U|V|W|Y|Z
<digit>::= 0|1|2|3|4|5|6|7|8|9 (Note:each of these appears
                                    in  every look-up table).
<identifier list>::= <identifier>|<identifier list>,<identifier>
```

2.   Numbers (1.3.1)

```
<real number>::= <unscaled real>
<unscaled real>::=<integer number>.<integer number>|
                                    .<integer number>
```

```
<integer number>::=<digit>|<integer number><digit>
```

3.   Declarations (1.4)

```
<declaration>::=<simple variable declaration>|
                <t array declaration>
```

3.1   Simple Variable Declarations (1.4.1)

```
<simple variable declaration>::=<simple type><identifier list>
<simple type>::= INTEGER|REAL
```

3.2   Array Declarations (1.4.2)

```
<t array declaration>::=<simple type>ARRAY<identifier list>
                                     (<bound pair list>)
<bound pair list>::=<bound pair>
<bound pair>::=<lower bound>::<upper bound>
<lower bound>::=<integer expression>
<upper bound>::=<integer expression>
```

4.   Expressions (1.5)

```
<t expression>::=<simple t expression>
```

4.1 Variables (1.5.1)

```
<simple t variable>::= <t variable identifier>|
                 <t array designator>
<t variable>::=<simple t variable>
<t array designator>::=<t array identifier>(<subscript list>)
<subscript list>::=<subscript>
<subscript>::=<integer expression>
```

4.2   Arithmetic Expressions (1.5.3)

```
<simple t expression>::=<t term>|<simple t expression>+
       <t term>|<simple t expression>-<t term>
<t term>::=<t factor>|<t term>*<t factor>
<t factor>::=<t primary>
<t primary>::=<t variable>|<t number>
```

4.3   Logical Expressions (1.5.4)

```
<logical expression>::=<relation>
<relation>::=<simple t expression><relational operator>
                              <simple t expression>
<relational operator>::= <
```

5.   Statements (1.6)

```
<program>::=<block>.<data>   (Note we do not provide a
            specification of the syntax of <data>).
<statement>::=<simple statement>|
          <iterative statement>|<if statement>
<simple statement> ::=<block>|<t assignment statement>|
                  <standard procedure statement>
```

5.1   Blocks (1.6.1)

```
<block>::=<block body><statement>END
<block body>::=<block head>|<block body><statement>;
<block head>::= BEGIN|<block head><declaration>
```

5.2   Assignment Statements (1.6.2)

```
<t assignment statement>::=<t left part><t expression>
```

<t left part>::=<t variable>:=

5.3   Standard Procedure Statements (cf. 1.6.3 and 1.6.8)

<standard procedure statement>::=<standard procedure identifier>

(<actual parameter list>)

<actual parameter list>::=<actual parameter>|

<actual parameter list>,<actual parameter>

<actual parameter>::=<t expression>|<t subarray designator>

<t subarray designator>::=<t array identifier>(<subarray

designator list>)

<subarray designator list>::=<subscript>

5.4   If Statements (1.6.5)

<if statement>::=<if clause><simple statement>

ELSE<statement>

<if clause>::= IF<logical expression>THEN

5.5   Iterative Statements (1.6.7)

<iterative statement>::=<for clause><statement>

<for clause>::= FOR<identifier>:=<initial value>UNTIL

<limit>DO

<initial value>::=<integer expression>

<limit>::=<integer expression>

APPENDIX C

i	u_i	TABLE LOOK-UP FOR					
		MODEL I	II	III	IV	V	VI
		y_i	y_i'	n_i	e_i	a_i	w_i
1	-.109	53.8	.3	53.8	53.8	53.8	53.6
2	0	53.6	.1	53.6	53.6	53.6	53.6
3	.178	53.5	0	53.5	53.5	53.5	53.5
4	.339	53.5	0	53.5	53.5	53.5	53.5
5	.373	53.4	-.1	53.4	53.4	53.4	53.4
6	.441	53.1	-.4	-.2	-.2	53.1	53.1
7	.461	52.7	-.8	-.4	-.2	52.7	52.7
8	.348	52.4	-1.1	-.4	-.2	.1	1
9	.127	52.2	-1.3	-.3	-.1	0	.1
10	-.18	52.0	-1.5	-.2	-.1	-.1	-.1
11	-.588	52.0	-1.5	-.1	0	.1	.1
12	-1.055	52.4	-1.1	.2	.3	.2	.2
13	-1.421	53.0	-.5	.4	.3	-.1	.1
14	-1.52	54.0	.5	.8	.5	.2	.2
15	-1.302	54.9	1.4	.8	.3	-.3	-.2
16	-.814	56.0	2.5	.8	.3	.2	.1
17	-.475	56.8	3.3	.5	.1	-.2	-.2
18	-.193	56.8	3.3	-.2	-.5	-.4	-.5
19	.088	56.4	2.9	-.8	-.6	.2	-.1
20	.435	55.7	2.2	-1.0	-.6	0	0
21	.771	55.0	1.5	-1.0	-.4	.2	.1
22	.866	54.3	.8	-.9	-.3	-.1	0
23	.875	53.2	-.3	-1.0	-.5	-.4	-.4
24	.891	52.3	-1.2	-1.0	-.4	.3	.1
25	.987	51.6	-1.9	-.7	-.2	0	.1
26	1.263	51.2	-2.3	-.4	0	.1	.1
27	1.775	50.8	-2.7	-.3	-.1	-.2	-.2
28	1.976	50.5	-3.0	-.2	-.1	.1	0

i	u_i	y_i	y_i	n_i	e_i	a_i	w_i
29	1.934	50.0	−3.5	−.4	−.2	−.1	−.2
30	1.866	49.2	−4.3	−.5	−.3	0	−.1
31	1.832	48.4	−5.1	−.5	−.2	.1	0
32	1.767	47.9	−5.6	−.2	.1	.2	.2
33	1.608	47.6	−5.9	0	.1	−.1	0
34	1.265	47.5	−6.0	.1	.2	0	0
35	.79	47.5	−6.0	.2	.1	0	0
36	.36	47.6	−5.9	.2	.1	0	0
37	.115	48.1	−5.4	.4	.3	.2	.2
38	.088	49.0	−4.5	.6	.4	.1	.2
39	.331	50.0	−3.5	.7	.3	−.2	−.1
40	.645	51.1	−2.4	.7	.3	.1	0
41	.96	51.8	−1.7	.4	0	−.2	−.2
42	1.409	51.9	−1.6	0	−.3	−.1	−.3
43	2.67	51.7	−1.8	−.3	−.3	.2	0
44	2.834	51.2	−2.3	−.4	−.3	0	0
45	2.812	50.0	−3.5	−.9	−.7	−.4	−.4
46	2.483	48.3	−5.2	−1.2	−.7	.2	−.1
47	1.929	47.0	−6.5	−.9	−.2	.4	.3
48	1.485	45.8	−7.7	−.5	0	−.1	.1
49	1.214	45.6	−7.9	.1	.4	.2	.3
50	1.239	46.0	−7.5	.6	.5	−.1	.1
51	1.608	46.9	−6.6	.9	.6	.1	.1
52	1.905	47.8	−5.7	.9	.4	−.2	−.1
53	2.023	48.2	−5.3	.5	−.1	−.3	−.4
54	1.815	48.3	−5.2	.2	−.1	.2	0
55	.535	47.9	−5.6	−.2	−.3	−.1	−.1
56	.122	47.2	−6.3	−.5	−.4	0	−.1
57	.009	47.2	−6.3	−.2	.1	.5	.4
58	.164	48.1	−5.4	.2	.3	−.1	.2
59	.671	49.4	−4.1	.4	.3	−.1	0
60	1.019	50.6	−2.9	.1	−.1	−.4	−.4
61	1.146	51.5	−2.0	0	−.1	.3	.1
62	1.155	51.6	−1.9	−.3	−.2	−.2	−.2

i	u_i	y_i	y_i'	n_i	e_i	a_i	w_i
63	1.112	51.2	-2.3	-.4	-.3	.1	o
64	1.121	50.5	-3.0	-.5	-.3	o	-.1
65	1.223	50.1	-3.4	-.3	o	.2	.2
66	1.257	49.8	-3.7	-.3	-.1	-.2	-.1
67	1.157	49.6	-3.9	-.3	-.1	o	-.1
68	.913	49.4	-4.1	-.3	-.1	o	-.1
69	.62	49.3	-4.2	-.2	-.1	.1	o
70	.255	49.2	-4.3	-.2	-.1	-.1	-.1
71	-.28	49.3	-4.2	-.2	-.1	o	o
72	-1.08	49.7	-3.8	-.2	o	o	o
73	-1.551	50.3	-3.2	-.2	-.1	-.1	-.1
74	-1.799	51.3	-2.2	-.1	o	.1	.1
75	-1.825	52.8	- .7	o	.1	.1	.1
76	-1.456	54.4	.9	o	o	⁊.1	-.1
77	-.944	56.0	2.5	o	o	o	o
78	-.57	56.9	3.4	-.4	-.4	-.3	-.4
79	-.431	57.5	4.0	-.5	-.3	.3	.1
80	-.577	57.3	3.8	-.7	-.4	-.3	-.2
81	-.96	56.6	3.1	-.8	-.4	o	-.1
82	-1.616	56.0	2.5	-.6	-.2	.2	.2
83	-1.875	55.4	1.9	-.6	-.2	-.3	-.2
84	-1.891	55.4	1.9	-.4	-.1	.2	.1
85	-1.746	56.4	2.9	.1	.4	.3	.4
86	-1.474	57.2	3.7	.1	o	-.6	-.4
87	-1.201	58.0	4.5	-.1	-.1	.1	-.1
88	-.927	58.4	4.9	-.2	-.2	-.1	-.1
89	-.524	58.4	4.9	-.4	-.3	o	o
90	.04	58.1	4.6	-.5	-.2	o	o
91	.788	57.7	4.2	.4	-.1	.1	.1
92	.943	57.0	3.5	-.3	-.1	-.1	-.1
93	.93	56.0	2.5	-.3	-.1	o	o
94	1.006	54.7	1.2	-.2	o	.1	.1
95	1.137	53.2	-.3	-.3	-.1	-.2	-.1
96	1.198	52.1	-1.4	-.1	.1	.3	.2

i	u_i	y_i	y_i	n_i	e_i	a_i	w_i
97	1.054	51.6	-1.9	.3	.3	.1	.2
98	.595	51.0	-2.5	.3	.1	-.3	-.2
99	-.08	50.5	-3.0	.2	.1	.1	0
100	-.314	50.4	-3.1	.4	.3	.3	.2
101	-.288	51.0	-2.5	.9	.7	.3	.4
102	-.153	51.8	-1.7	1.1	.5	-.3	-.1
103	-.109	52.4	-1.1	.7	.1	-.3	-.4
104	-.187	53.0	-.5	.3	-.1	.1	-.1
105	-.255	53.4	-.1	.1	-.1	.1	.1
106	-.229	53.6	.1	0	0	0	.1
107	-.007	53.7	.2	0	0	0	0
108	.254	53.8	.3	0	0	0	Q
109	.33	53.8	.3	-.1	-.1	-.1	-.1
110	.102	53.8	.3	-.1	0	.1	.1
111	-.423	53.3	-.2	-.3	-.3	-.3	-.3
112	-1.139	53.0	-.5	-.3	-.1	.4	.2
113	-2.275	52.9	-.6	-.1	.1	0	.1
114	-2.594	53.4	-.1	.2	.3	.2	.2
115	-2.716	54.6	1.1	.7	.5	.1	.2
116	-2.51	56.4	2.9	.9	.5	-.1	0
117	-1.79	58.0	4.5	.6	.1	-.3	-.3
118	-1.346	59.4	5.9	.2	-.2	0	-.2
119	-1.081	60.2	6.7	-.2	-.3	.1	-.1
120	-.91	60.0	6.5	-.7	-.6	-.2	-.3
121	-.876	59.4	5.9	-.9	-.5	.2	.1
122	-.885	58.4	4.9	-.9	-..4	-.1	-.1
123	-.8	57.6	4.1	-.8	-.3	.1	.1
124	-.544	56.9	3.4	-.7	-.3	-.2	-.1
125	-.416	56.4	2.9	-.7	-.3	0	-.1
126	-.271	56.0	2.5	-.7	-.3	-.1	-.1
127	0	55.7	2.2	-.6	-.2	.1	C
128	.403	55.3	1.8	-.6	-.2	-.1	-.1
129	.841	55.0	1.5	-.4	-.1	.2	.1
130	1.285	54.4	.9	-.5	-.2	-.3	-.2
131	1.607	53.7	.2	-.5	-.2	.1	C

i	u_i	y_i	y_i	n_i	e_i	a_i	w_i
132	1.746	52.8	-.7	-.4	-.2	0	0
133	1.683	51.6	-1.9	-.5	-.2	-.1	-.1
134	1.485	50.6	-2.9	-.3	0	.3	.2
135	.993	49.4	-4.1	-.3	-.2	-.4	-.2
136	.648	48.8	-4.7	-.1	.1	.4	.2
137	.577	48.5	-5·0	0	.1	-.2	0
138	.577	48.7	-4.8	.1	.1	0	0
139	.632	49.2	-4.3	.1	0	-.1	-.1
140	.747	49.8	-3.7	0	-.1	-.1	-.1
141	.9	50.4	-3.1	-.1	0	.1	0
142	.993	50.7	-2.8	-.1	-.1	-.1	-.1
143	.968	50.9	-2.6	-.1	0	.1	.1
144	.79	50.7	-2.8	-.2	-.1	-.2	-.2
145	.399	50.5	-3·0	-.2	-.1	.2	.1
146	-.161	50.4	-3.1	-.1	0	.1	.1
147	-.553	50.2	-3.3	-.2	-.2	-.3	-.2
148	-.603	50.4	-3.1	-.2	-.1	.1	0
149	-.424	51.2	-2.3	-.1	0	.1	.1
150	-.194	52.3	-1.2	0	.1	-.1	0
151	-.049	53.2	-.3	-.1	-.1	-.2	-.2
152	.06	53.9	.4	-.2	-.1	.2	.1
153	.161	54.1	.6	-.2	-.1	-.1	-.1
154	.301	54·0	.5	-.2	-.1	0	0
155	.517	53.6	.1	-.3	-.2	-.1	-.1
156	.566	53.2	-.3	-.4	-.2	0	0
157	.56	53·0	-.5	-.3	0	.2	.2
158	.573	52.8	-.7	0	.1	0	.1
159	.592	52.3	-1.2	-.1	-.1	-.3	-.2
160	.671	51.9	-1.6	-.1	-.1	.1	0
161	.933	51.6	-1.9	-.2	-.1	-.1	-.1
162	1.337	51.6	-1.9	0	.1	.2	.2
163	1.46	51.4	-2.1	-.1	-.1	-.3	-.2
164	1.353	51.2	-2.3	0	0	.2	.1

i	u_i	y_i	y_i'	n_i	e_i	a_i	w_i
165	.772	50.7	-2.8	0	0	-.1	0
166	.218	50.0	-3.5	-.1	-.1	0	-.1
167	-.237	49.4	-4.1	-.1	-.1	0	0
168	-.714	49.3	-4.2	-.2	-.1	0	-.1
169	-1.099	49.7	-3.8	-.3	-.2	-.1	-.1
170	-1.269	50.6	-2.9	-.5	-.3	0	-.1
171	-1.175	51.8	-1.7	-.6	-.3	0	-.1
172	-.676	53.0	-.5	-.7	-.4	-.1	-.2
173	.033	54.0	.5	-1.0	-.6	-.1	-.2
174	.556	55.3	1.8	-.6	-.1	.6	.4
175	.643	55.9	2.4	-.4	0	-.3	0
176	.484	55.9	2.4	0	.2	.2	.2
177	.109	54.6	1.1	-.3	-.3	-.6	-.5
178	-.31	53.5	0	-.2	0	.5	.2
179	-.697	52.4	-1.1	-.4	-.3	-.4	-.3
180	-1.047	52.1	-1.4	-.4	-.1	.3	.1
181	-1.218	52.3	-1.2	-.4	-.2	-.2	-.1
182	-1.183	53.0	-.5	-.4	-.2	.1	0
183	-.873	53.8	.3	-.5	-.3	-.2	-.2
184	-.336	54.6	1.1	-.7	-.4	0	-.1
185	.063	55.4	1.9	-.7	-.3	.1	.1
186	.084	55.9	2.4	-.5	-.1	0	.1
187	0	55.9	2.4	-.3	0	0	.1
188	.001	55.2	1.7	-.4	-.2	-.2	-.2
189	.209	54.4	.9	-.3	-.1	.1	0
190	.556	53.7	.2	-.4	-.2	-.1	-.1
191	.782	53.6	.1	-.1	.1	.3	.2
192	.858	53.6	.1	.1	.2	0	.1
193	.918	53.2	-.3	.2	.1	-.2	-.1
194	.862	52.5	-1.0	.1	0	0	-.1
195	.416	52.0	-1.5	.2	.2	.3	.2
196	-.336	51.4	-2.1	.1	0	-.2	-.1
197	-.959	51.0	-2.5	.1	0	.1	0
198	-1.813	50.9	-2.6	0	-.1	-.1	-.1

i	u_i	y_i	y_i'	n_i	e_i	a_i	w_i
199	-2.378	52.4	-1.1	.9	.9	1.0	1.0
200	-2.499	53.5	0	.8	.3	-1.1	-.6
201	-2.473	55.6	2.1	1.2	.7	.8	.5
202	-2.33	58.0	4.5	1.6	.9	.1	.4
203	-.053	59.5	6.0	1.3	.4	-.7	-.4
204	-1.739	60.0	6.5	.4	-.3	-.3	-.5
205	-1.261	60.4	6.9	0	-.2	.5	.2
206	-.569	60.5	7.0	.9	.9	1.0	1.2.
207	-.137	60.2	6.7	1.1	.6	-1.0	-.3
208	-.024	59.7	6.2	1.7	1.0	.7	.6
209	-.05	59.0	5.5	1.3	.3	-.9	-.6
210	-.135	57.6	4.1	.8	0	.2	-.1
211	-.276	56.4	2.9	.7	.2	.4	.3
212	-.534	55.2	1.7	.4	0	-.3	-.2
213	-.871	54.5	1.0	.2	0	.2	.1
214	-1.243	54.1	.6	0	-.1	-.1	-.1
215	-1.439	54.1	.6	-.2	-.1	.1	0
216	-1.422	54.4	.9	-.3	-.2	-.1	0
217	-1.175	55.5	2.0	.1	.3	.5	.5
218	-.813	56.2	2.7	0	0	-.6	-.3
219	-.634	57.0	3.5	.1	.1	.3	.2
220	-.582	57.3	3.8	0	0	-.2	-.1
221	-.625	57.4	3.9	.2	.2	.3	.2
222	-.713	57.0	3.5	.2	.1	-.2	-.1
223	-.848	56.4	2.9	.1	0	0	-.1
224	-1.039	55.9	2.4	0	-.1	0	0
225	-1.346	55.5	2.0	-.3	-.2	-.1	-.1
226	-1.628	55.3	1.8	-.5	-.3	-.2	-.1
227	-1.619	55.2	1.7	-.8	-.5	0	-.2
228	-1.149	55.4	1.9	-1.0	-.6	.1	-.1
229	-.488	56.0	2.5	-1.0	-.4	-.2	0
230	-.160	56.5	3.0	-1.1	-.5	.4	-.2
231	-.007	57.1	3.6	-.7	-.1	.1	.3
232	-.092	57.3	3.8	-.2	.3	-.2	.3

i	u_i	y_i	y_i'	n_i	e_i	a_i	w_i
233	-.62	56.8	3.3	.2	.3	-.2	0
234	-1.086	55.6	2.1	.1	-.1	-.4	-.4
235	-1.525	55.0	1.5	.2	.2	.5	.3
236	-1.585	54.1	.6	-.4	-.6	-.9	-.8
237	-2.029	54.3	.8	-.6	-.3	.7	.2
238	-2.024	55.3	1.8	-.5	-.1	0	.2
239	-1.961	56.4	2.9	-.3	0	-.1	0
240	-1.952	57.2	3.7	-.5	-.3	-.4	-.3
241	-1.794	57.8	4.3	-.7	-.4	.1	-.1
242	-1.302	58.3	4.8	-.9	-.4	0	-.1
243	-1.03	58.6	5.1	-.9	-.4	0	0
244	-.918	58.8	5.3	-.8	-.3	.1	.1
245	-.798	58.8	5.3	-.5	-.1	.1	.1
246	-.867	58.6	5.1	-.1	.1	.1	.2
247	-1.047	58.0	4.5	0	.1	-.2	-.1
248	-1.123	57.4	3.9	.1	.1	0	0
249	-.876	57.0	3.5	.1	.1	0	0
250	-.395	56.4	2.9	-.3	-.4	-.4	-.4
251	.185	56.3	2.8	-.4	-.2	.4	.1
252	.662	56.4	2.9	-.3	-.1	.1	.1
253	.709	56.4	2.9	0	.2	.1	.2
254	.605	56.0	2.5	.4	.4	.1	.2
255	.501	55.2	1.7	.7	.5	0	.2
256	.603	54.0	.5	.7	.3	-.3	-.2
257	.943	53.0	-.5	.6	.2	.1	0
258	1.223	52.0	-1.5	0	-.3	-.4	-.4
259	1.249	51.6	-1.9	-.1	-.2	.4	.2
260	.824	51.6	-1.9	.2	.2	.3	.4
261	.102	51.1	-2.4	.2	.1	-.4	-.2
262	.025	50.4	-3.1	0	-.1	0	-.1
263	.382	50.0	-3.5	-.1	-.1	0	-.1
264	.922	50.0	-3.5	-.5	-.4	-.3	-.3
265	1.032	52.0	-1.5	.8	1.1	1.6	1.4
266	.866	54.0	.5	2.2	1.7	-.2	.7

i	u_i	y_i	y_i'	n_i	e_i	a_i	w_i
267	.527	55.1	1.6	3.3	2.1	.1	.5
268	.093	54.5	1.0	3.1	1.2	-.8	-.5
269	-.458	52.8	-.7	1.9	.1	-.5	-.8
270	-.748	51.4	-2.1	.6	-.5	.1	-.4
271	-.947	50.8	-2.7	-.4	-.7	.1	-.1
272	-1.029	51.2	-2.3	-.8	-.6	.2	.1
273	-.928	52.0	-1.5	-1.1	-.6	-.1	-.1
274	-.645	52.8	-.7	-1.4	-.8	-.2	-.2
275	-.424	53.8	.3	-1.3	-.5	.3	.2
276	-.276	54.5	1.0	-1.2	-.5	-.2	-.1
277	-.158	54.9	1.4	-1.0	-.3	.1	0
278	-.033	54.9	1.4	-.9	-.3	-.1	-.1
279	.102	54.8	1.3	-.6	-.1	.1	.1
280	.251	54.4	.9	-.5	-.2	-.2	-.2
281	.28	53.7	.2	-.8	-.5	-.2	-.3
282	0	53.3	-.2	-.7	-.3	.3	.1
283	-.493	52.8	-.7	-.8	-.4	-.2	-.2
284	-.759	52.6	-.9	-.6	-.1	.2	.2
285	-.824	52.6	-.9	-.4	-.1	-.1	0
286	-.74	53.0	-.5	-.3	0	0	0
287	-.528	54.3	.8	.4	.6	.6	.6
288	-.204	56.0	2.5	1.4	1.1	.2	.6
289	.034	57.0	3.5	1.8	1.0	-.3	0
290	.204	58.0	4.5	2.6	1.6	.7	.7
291	.253	58.6	5.1	3.4	1.9	.1	.5
292	.195	58.5	5.0	3 8	1.8	-.1	.2
293	.131	58.3	4.8	4.2	2.0	.4	.6
294	.017	57.8	4.3	4.2	1.8	-.1	.2
295	-.182	57.3	3.8	4.1	1.7	.2	.3
296	-.262	57.0	3.5	4.0	1.6	.2	.4

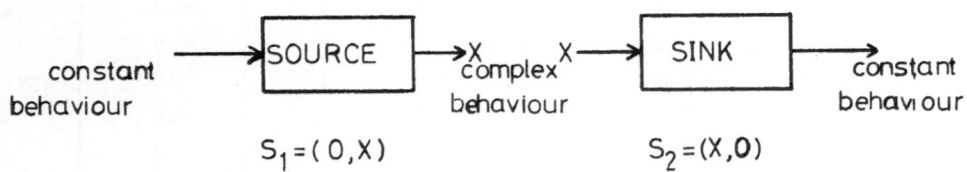

constant behaviour → SOURCE →X complex X → SINK → constant behaviour

$S_1 = (0, X)$ $S_2 = (X, 0)$

FIG   1

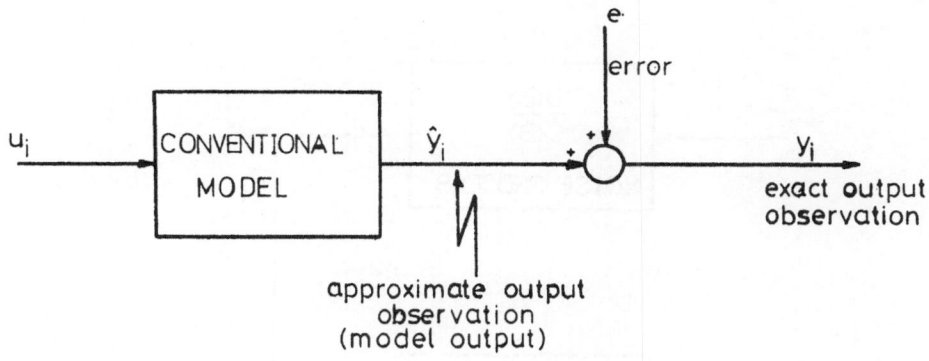

Representation of Conventional Model

FIG 2a

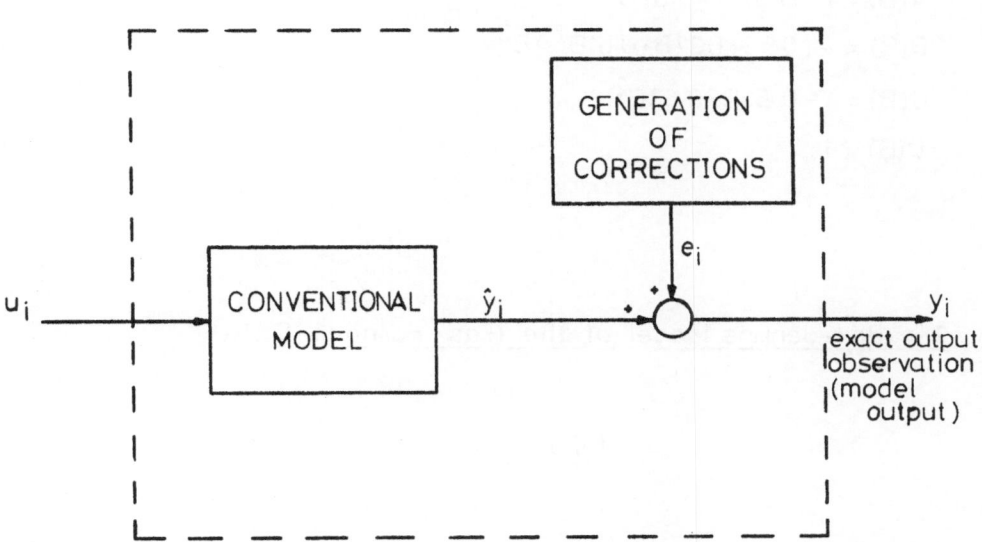

Corresponding Model Which Satisfies Definitions 3.3.1 , 3.3.6

FIG 2b

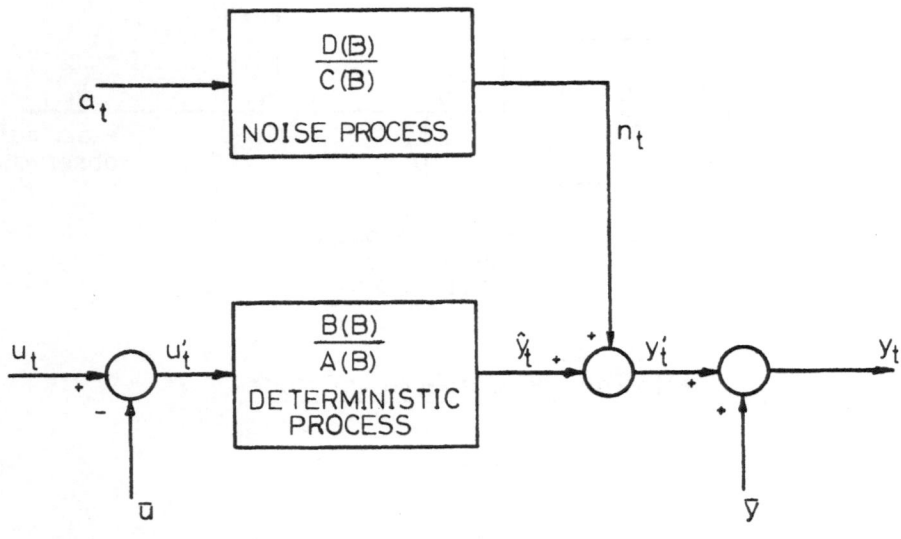

$A(B) = 1 - 0.57B - 0.01B^2$

$B(B) = -(0.53 + 0.37B + 0.51B^2)B^3$

$C(B) = 1 - 0.53B + 0.63B^2$

$D(B) = 1$

The Box-Jenkins Model of the Gas-Furnace Data

FIG 3

(a) Model I – The Trivial Model

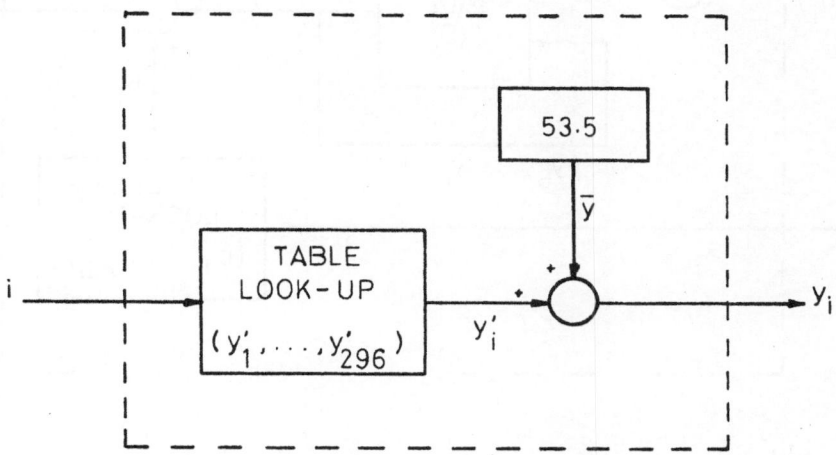

(b) Model II – The Mean

FIG 4

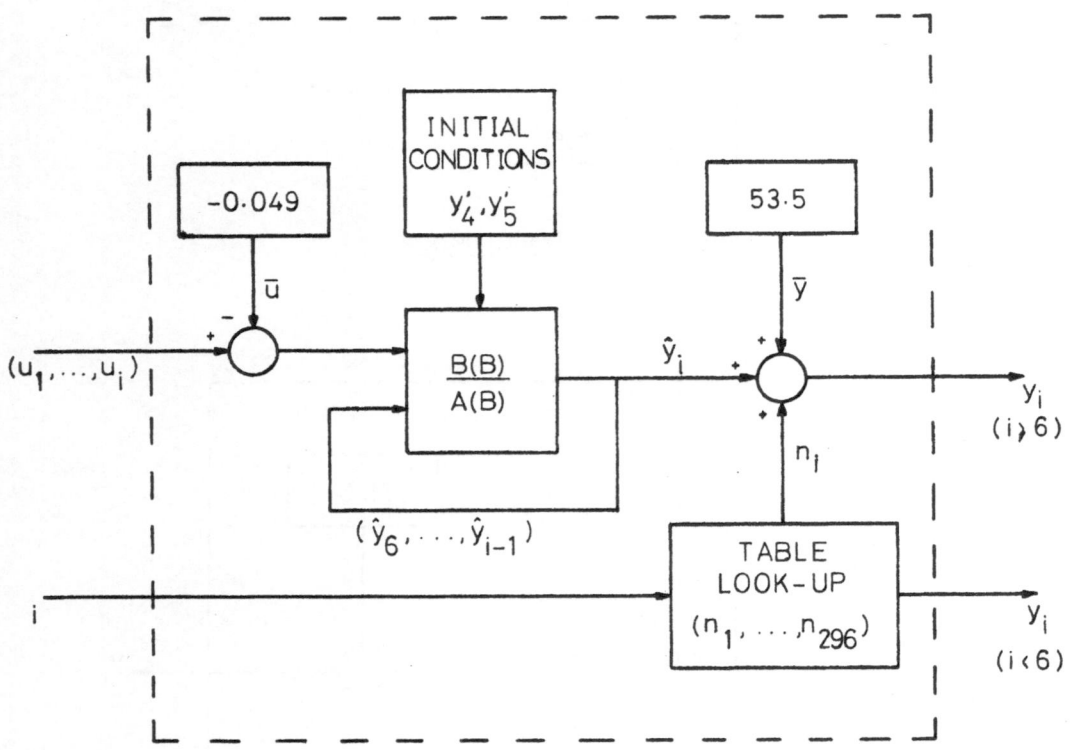

(c) Model III – Deterministic Transfer Function

Using Input Observations Only

FIG 4

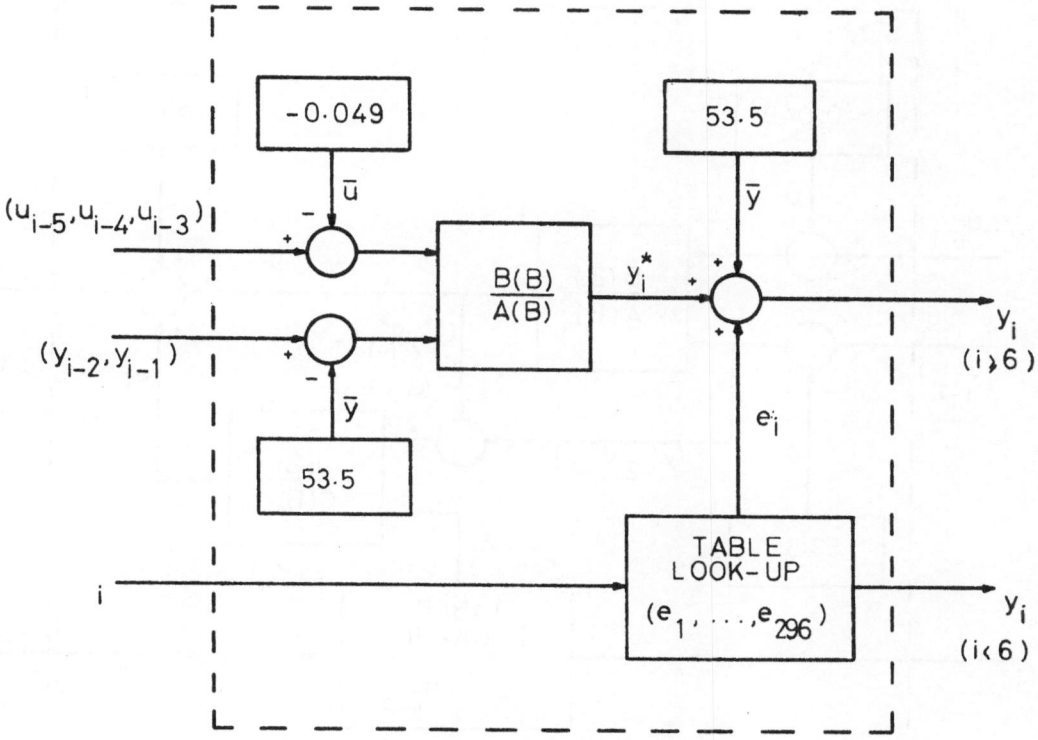

(d) Model IV – Deterministic Transfer Function

Using Input and Output Observations

FIG 4

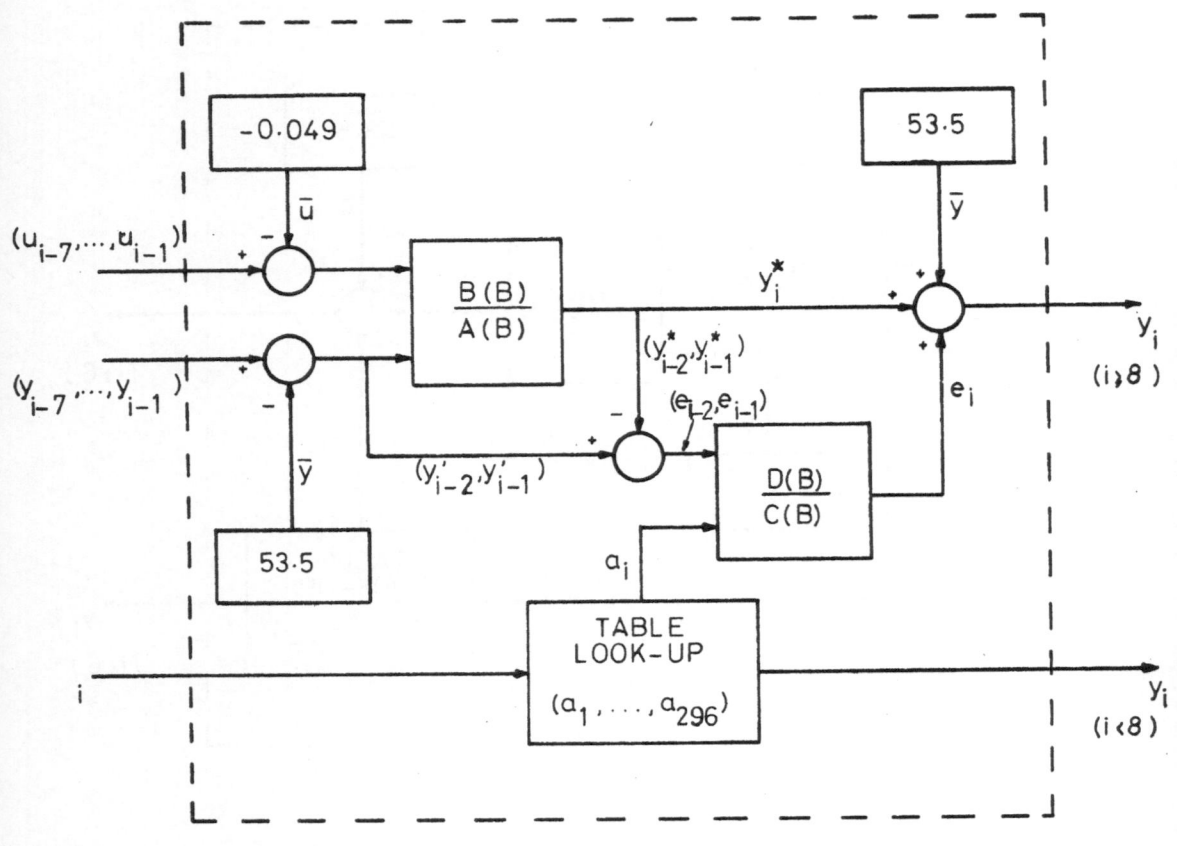

(e) Model V – Stochastic Process Model

FIG 4

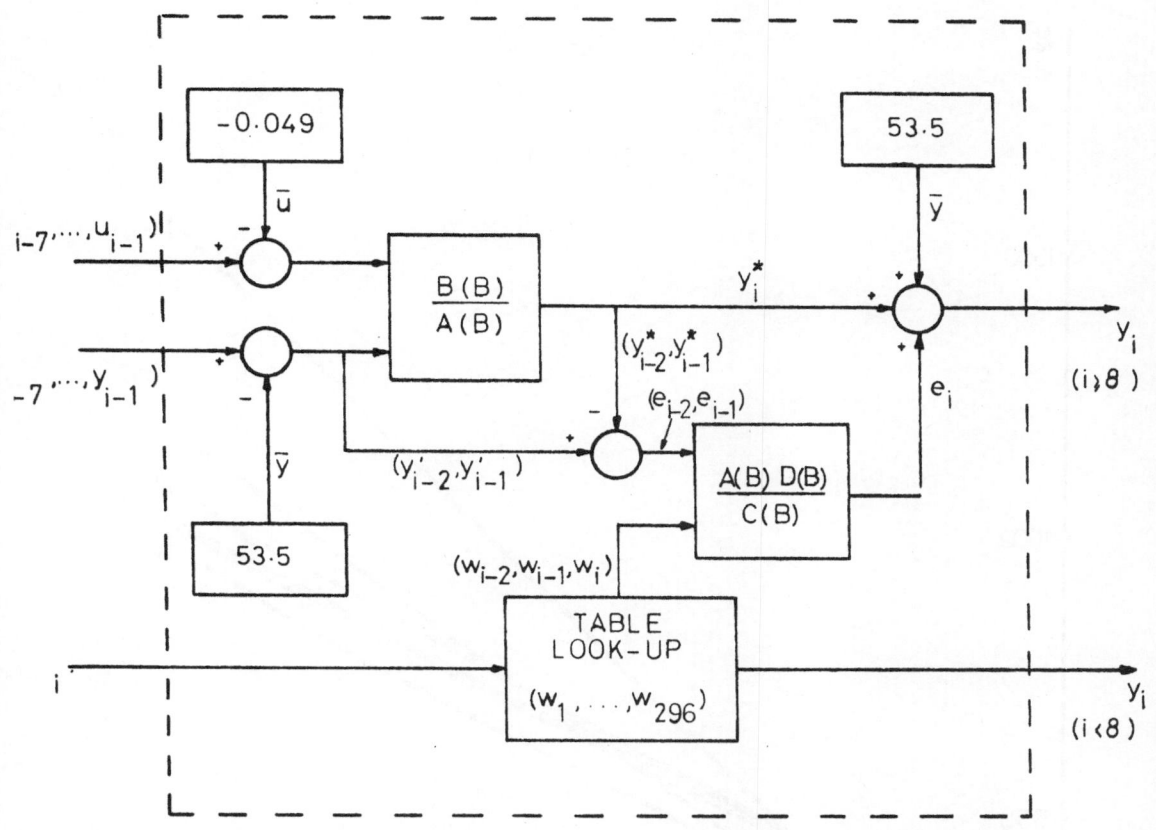

(f) Model VI- Box & Jenkins Model

FIG 4

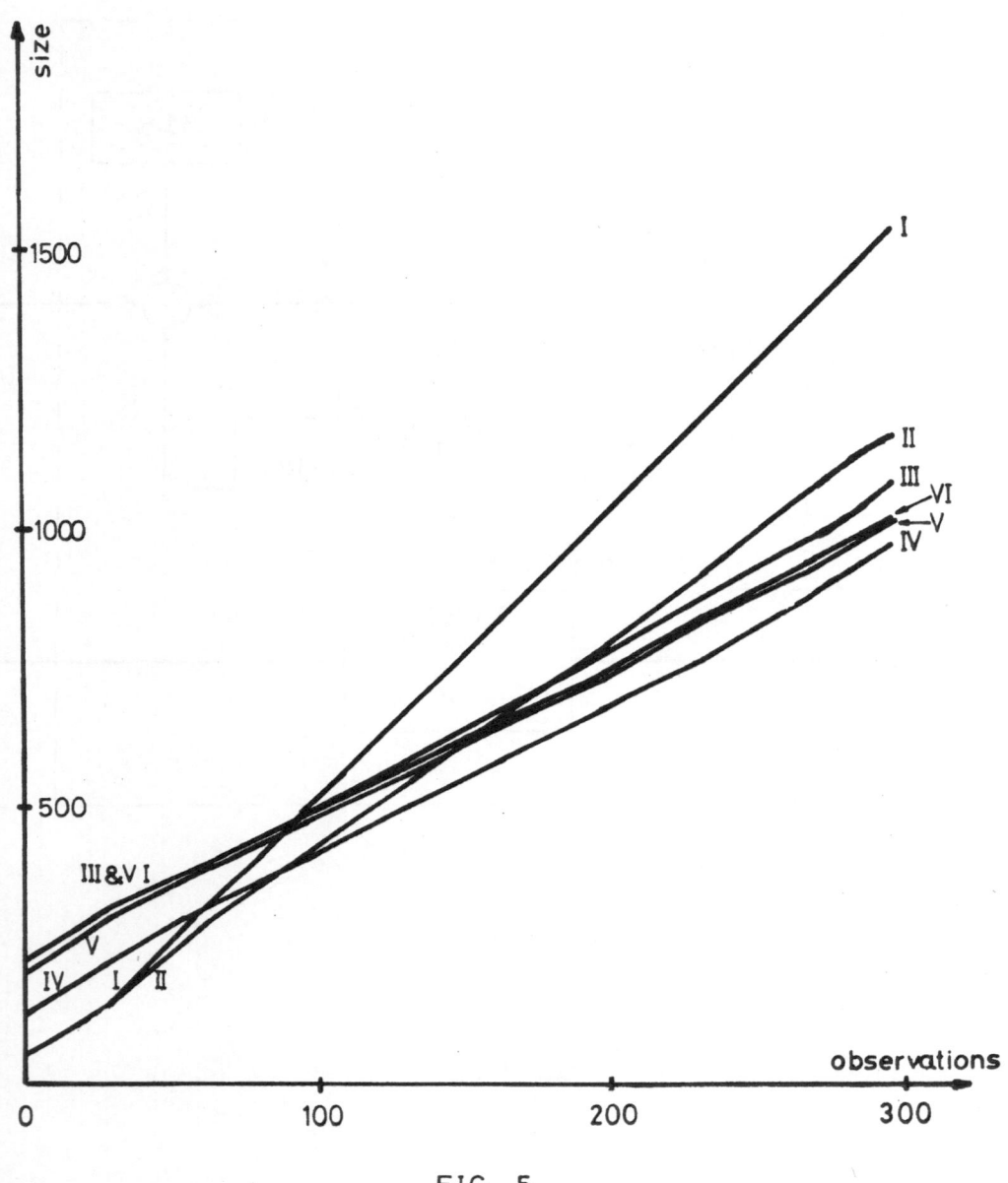

<u>FIG 5</u>

<u>Sizes of models I-VI</u>

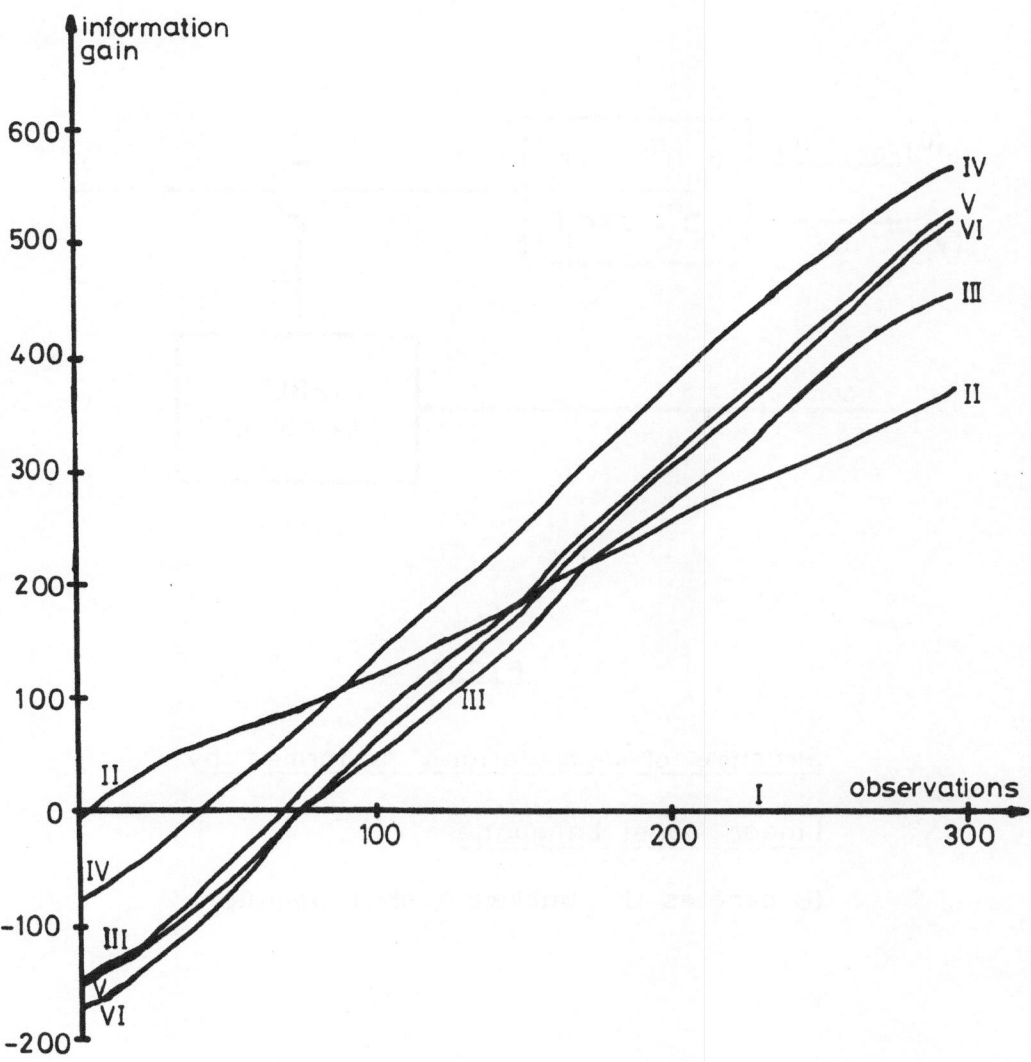

FIG 6

information gains of models I - VI

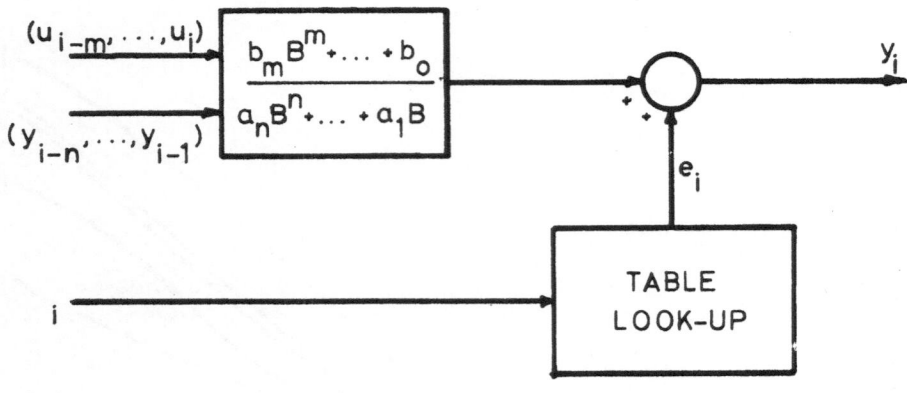

FIG 7

Structure of computations performed by Linear Model Language

(B denotes the backward shift operator)

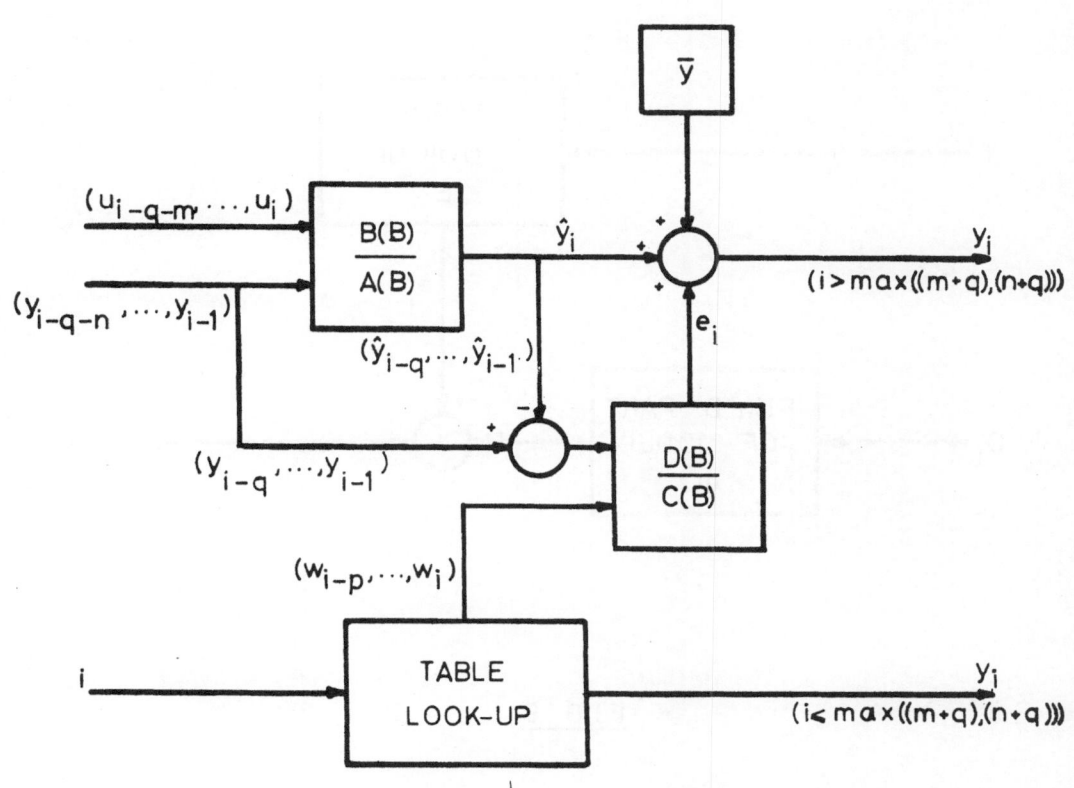

FIG 8

Structure of computations performed by Extended

Linear Model Language

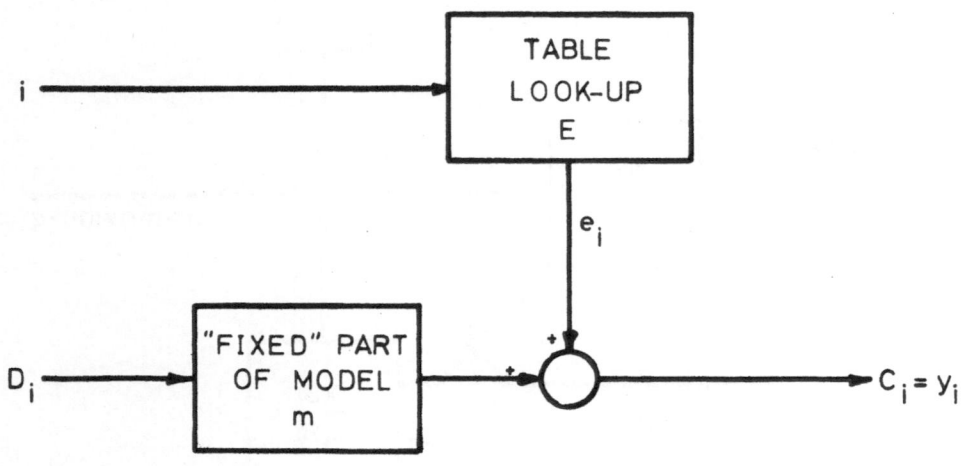

FIG 9

Assumed structure of models

(in chapter 7)

For definitions of D_i, C_i see def. (3.3.7)

Lecture Notes in Economics and Mathematical Systems